ENCYCLOPAEDIA OF PLANT PROTECTION

ENCYCLOPAEDIA OF PLANT PROTECTION

ENCYCLOPAEDIA OF
PLANT PROTECTION

Vol. 4

B.A. Mannan

ANMOL PUBLICATIONS PVT. LTD.
NEW DELHI - 110 002 (INDIA)

ANMOL PUBLICATIONS PVT. LTD.
H.O.: 4374/4B, Ansari Road, Darya Ganj,
New Delhi-110 002 (India)
Ph.: 23278000, 23261597
B.O.: No. 1015, Ist Main Road, BSK IIIrd Stage
IIIrd Phase, IIIrd Block,
Bangalore - 560 085 (India)
Visit us at: www.anmolpublications.com

Encyclopaedia of Plant Protection

First Published, 2008

ISBN 978-81-261-3538-7 (Set)

PRINTED IN INDIA

Printed at Mehra Offset Press, Delhi.

CONTENTS

PREFACE

Plant lovers, food producers and agricultural planners, all voice together to protect the plants from insects and diseases. Damages to the plants by insects are easily felt, but not realised fully. Every year, hundreds of people die of hunger. We can provide them with enough, food, if food crops are kept safe from the damages.

Plants, like human beings, are prone to different diseases. A diseased plant cannot yield good produces. Hence, plant protection has attained greater significance in today's world. Agricultural crops have been prone to damage, because of insects, rodents, diseases, weeds and nematodes, etc. since time immemorial.

Plant Protection continues to play a significant role in achieving targets of crop production. Major thrust areas of plant protection are promotion of integrated pest management, ensuring availability of safe and quality pesticides for sustaining crop production, despite the ravages of pests and diseases by streamlining the quarantine measures for accelerating the introduction of new high-yielding crop varieties, besides eliminating the chances of entry of exotic pests.

It is essential to have the basic knowledge about the plant pathogens, different kinds of diseases, caused by them to various crops and of course, the methods of controlling diseases to avoid enormous losses caused by the diseases. As plant protection is indispensable, job opportunities are always there for people, having basic knowledge of pests, crops and diseases.

This book is a serious endeavour in that direction. This *Encyclopaedia of Plant Protection,* in four volumes takes the stock of present state of art in plant protection. It also focuses on isolation, structure, characterisation and physiochemical and physiological properties of plants, synthesis, functions and mechanisms of action of plant protease inhibitors. It also deals with their inhibitory effects on insects and pathogens and with the resultant application of genetic engineering to plants for their protection.

This book caters to the need of knowledge hungry biologists, agriculturists, policy planners and specialists in the field of plant protection and of course the students, teachers, researchers and beyond that, all those concerned with agricultural production and its development.

Editor

1

Controlling Devices

The use of pesticides has increased considerably throughout the world over during the last three decades. With the need for increased production without increasing the area of crops, more farmers are using improved varieties of seeds, fertilisers and plant protection chemicals. Pest and disease control is often essential. When we use pesticide, proper handling and care is essential otherwise it leads hazards to human beings and environment.

Pesticides may get entry into body by three routes namely by mouth (oral), through the lungs (inhalation) or through the skin (dermal). Poisoning may result from a single dose of pesticide which referred to as "acute" poisoning or by repeated in take of small quantities of poison *i.e.*, chronic poisoning.

Based on acute toxicity data, usually for rats and rabbits, pesticides are often grouped into arbitrary categories of hazard and for case of reference, an example of this type of classification is given in table.

Level of Hazard	*Pesticides*	*LD 50 in rats mg/kg of body weight*		*Colour of Triangle Container*
		Oral	*Dermal*	
Extremely Hazardous	Temik, Endrin Thimet Dimecron, Dieldriin, TEP	Less than 50	Less than 200	Bright red
Very Hazardous	Aldrin, Fenthion Toxaphene, DDVP DDT, Thiodan Rogor yellow Sumithion, Chlordane, Lindane Diazinon	51-500	201-2000	Bright
Moderately Hazardous	Sevin, Malathion, Pyrethrin, Orthene	501-5000	2001-20,000	Bright Blue
Relatively Safe	Sulphur, Methoxychlor	More than 15,000	More than 120,000	Bright Green

Safety precautions should be taken when applying pesticides of different categories of hazard. Improper handling of pesticides may result in fatality or lethal effect to a person. Protective clothing, rubber gloves, shoes and large handkerchief are very much essential to protect the body, hands, legs and face.

First aid should be given when a person using pesticides becomes ill. General symptoms of poisoning are headache, nausea, vomiting, anxiety and in severe case convulsions occur. Immediate treatment such as washing the contaminated skin, flushing the contaminated eye, artificial respiration and vomiting induced with saline solution must be employed for immediate recovery from poison. Immediate medical attention by a Doctor is essential. The doctor must be informed the name of the pesticide and given as much information as possible by showing him a leaflet or label about the chemical so that he can administer effective antidote for immediate recovery.

Other safety precautions for pesticides application are given below:

Before Applying Pesticide-General Instructions

1. Know the pest, and it's extent of damage.
2. Use pesticide only when really needed.
3. Seek advice on the proper method of control.
4. Use only the recommended pesticide for the problem. If several pesticides are recommended, choose the least toxic to mammals and if possible the least persistent.
5. Read the label inducing the small print.
6. Make sure that appropriate protective clothing is available and is used.
7. Commercial operators using large quantities of organophosphate pesticides should visit their doctor and get their blood tested.
8. Check application equipment for leaks, and ensure that it is in proper working order.
9. Check that pesticides are properly placed in locked store. Avoid inhaling pesticide mists or dusts, specially in confined spaces such as the pesticide store.
10. Warn neighbours about your spray programme, specially if they have apiaries (Honey bees).
11. Take only required pesticide for the application from the store to the site of application. Do not transfer pesticides into other containers.

While Mixing and Applying Pesticides

1. Wear appropriate protective clothing. If it is contaminated remove and replace with clean clothing.
2. Never work alone when handling the most toxic pesticides.

3. Never allow children or other unauthorised persons for this job.
4. Re-check the instructions on the label.
5. Avoid contamination of the skin, specially the eyes and mouth. Formulations should be poured carefully to avoid splashing. Avoid powder formulations puffing up into the face. If contaminated with the concentrate, wash immediately.
6. Never eat, drink or smoke while mixing or applying pesticides.
7. Always have plenty of water available for washing.
8. Always stand upwind while mixing pesticide.
9. Make sure that pesticides are mixed in the correct quantities.
10. Avoid inhalation of chemical, dust or fumes.
11. Start spraying near the downwind edge of the field and proceed upwind so that operators move into unsprayed areas.
12. Never blow out clogged nozzles or hoses with your mouth.
13. Avoid spray drift and do not spray if wind conditions cause drift, otherwise birds, bees and wildlife may be endangered. Never spray if the wind is blowing towards grazing livestock or regularly used pastures.
14. Never leave pesticides unattended in field.
15. Provide proper supervision of those assisting with pesticides application and have adequate rest periods.
16. Do not work with pesticides if your cholinestrase level is below normal.

Instructions after Application

1. Return unused pesticide to the store.

2. Safely dispose of all empty containers. Since it may be difficult to bury empty containers after each day's spraying operations, hence they should be kept in the pesticide store until a convenient number is ready for disposal. It is absolutely impossible to clean out a container sufficiently well to make it safe for use, for storage of food, water or as a cooking utensil. If any container is burnt, never stand in the smoke.
3. Never leave pesticide in application equipment. Clean *equipment and return it* to store.
4. Remove and clean the protective clothing.
5. Keep a record of the use of pesticides.
6. Do not allow persons *to enter the treated areas for the required period if* restrictions apply to the pesticide used.

Treatments in Case of Poisoning

In case of pesticidal poisoning to a person *following treatments should be followed* immediately:

1. The efforts should be made to remove the poison from the body of affected person.
2. Antidote should be administered as early as possible.
3. The excreta etc. should be collected and sent for further examination.
4. Arrangement should be made to carry patient to some hospital.

In Cases of Swallowed Poison

In case of swallowing the poison, efforts should be made for vomiting and stomach may be washed with stomach tube. For inducing vomiting, following substances may be given:

1. Saline solution (15 g Common salt in a glass of warm water).

2. One gram zinc sulphate in a glass of water.
3. Some quantity of soap in a glass of warm water.
4. 1/10 gram injection of apomorphine hydrochloride.

If the patient is unconscious, then vomiting should be avoided and help of the doctor must be taken immediately and after washing the stomach some antidote may be given. It is always better to give universal antidote which contains following ingredients:

(a) Charcoal — 2 parts
(b) Tamic acid — 1 part
(c) Milk of magnesia — 1 part

Some other Common Antidotes are as Follows

1. Atropine sulphate
2. BAM (Diacetyl monohexene)
3. PAM (Pyridine aldoxymet)

In Case of Inhalation of Poison

Following case should be observed in case of inhalating the poison:

1. The patient must be brought immediately in open air.
2. If he is in the house, all the ventilators and doors should be opened.
3. The cloths of the body may be loosen.
4. The patient must be saved from cold and may be wrapped with blanket.
5. If he is feeling trouble in respiration artificial oxygen may be employed.
6. In case of headache, convulsion and unconsciousness patient should be kept in dark places and avoid noise.

Antidotes for the Treatment of Pesticide Poisoning

Pesticide	*Symptoms of Poisoning*	*Antidotes*
Organochlorine		
Aldrin, chlordane, heptachlor, DDT, HCH, lindane, dicofor, endosulfan	Nervousness, headache, dizziness, nausea, vomiting, tremors and diarrhoea	(i) Phenobarbital 0.7 g per day or pentabarbital 0.5 g per day (ii) Do not give morphine, theophylline or aminophylline
Carbonates		
Carbaryl, aldicarb, thiobencarb, carbofuran, propoxur	Giddiness, headache, nausea, diarrhoea blurred vision muscular twitching, convulsions	(i) Inject atropine sulphate 2-4 mg. for an adult (0.04-0.08 mg/kg body weight for children for 24-48 hours. (ii) Convulsions and anxiety can be treated with 5.10 mg of diazepam injected intratnuscularly. (iii) While keeping the patient fully atropinised administer also an oxime, if available, *e.g.*, 2-PAM 1000-2000 mg IM or IV for adults or 15 mg/kg body weight for children. Repeat after 1-2 hours.
Organophosphates		
Malathion, methylparathion, acephate, monocrotophos	Same as carbonates	Same as carbonates
Nicotine		
	Neurological manifestations	Phenobarbitene sulphate 0.7 g daily
Arsenic Compounds		
	Severe gastrointestinal	BAL (Dimercal prol) 3-5 mg/kg body weight

The inorganic insecticides were widely used before the synthetic organic insecticides were developed and marketed. At present, hardly any of these insecticides are being used in India. There is a feeling in some quarters that we can perhaps advantageously use Arsenicals against the leaf-feeding insects, especially when some parasites or predators are known to be active against them.

Insecticides of plant origin have also been relegated to the background by the synthetics and at present relatively very little quantities of insecticides, like Pyrethrins and Nicotine sulphate, are used to control insect pests of agricultural importance. The problem of residues of insecticides on food and fodder crops has emphasized the necessity of using insecticides which are less toxic to mammals or breakdown into non-toxic components in a reasonably short time. Thus the insecticides of plant origin can be usefully employed to control a number of insect pests, especially where the products are to be consumed within a short time after treatment (vegetables and fruits).

The use of synthetic organic insecticides has increased with leaps and bounds throughout the world. In India, only about 10,120 hectares of crops ware treated with pesticides in 1946-47; this area increased to about 17.4 million hectares by 1965-66 and the target for 1973-74 was 80 million hectares. The various insecticides required for controlling insects during this period is estimated to be 40,850 tonnes.

Starting with the manufacturers of BHC in 1958, India today produces technical grades of over 39 pesticides (insecticides, fungicides, rodenticides, molluscicides, nematocides and herbicides), including the following insecticides used for controlling crop pests: BHC, DDT, Malathion, Parathion (Methyl), Toxaphene, Pyrethrum extract, Nicotine sulphate, Fenitrothion, Methyl-deneton, Phosphamidon and Dimethoate.

Classification of Machinery

Sprayers	*Dusters*	*Fumigators*
	Hand operated	*Power operated*
1. Automizer	1. Hydraulic power sprayer	1. Cyno gas pump
2. Slide pump		
3. Compressed air sprayer	2. Blower sprayer	2. Soil injector
4. Knap sack		
5. Foot sprayer		
6. Rocker sprayer	*Hand operated*	*Power operated*
7. Stirup sprayer	1. Plunger duster	1. Power duster
8. Charge pump	2. Blow duster	
	3. Rotatory duster	

Sprayers

The liquid formulation of insectides are used with the help of a sprayer and the qualities of a good sprayer are given below:

1. It should discharge uniform spray on the treated material.
2. The metal used for the construction of a sprayer should not react chemically with the insecticide and must be strong enough.
3. The construction should preferably be simpler so as to clean it without any difficulty.
4. Sprayer should treat maximum area in minimum time.
5. Solution must be broken down into very fine uniform droplets before it reaches the materials to be treated.
6. Spray must cover the desired distance and height.
7. There should not be any wastage of chemical either through leakage or falling from the sprayer when using it in field.

The description of some commonly used hand operated sprayers is given below:

Hand Automizer (Flit Pump)

It is generally used to kill mosquitoes, houseflies etc. by spraying the chemical in houses. This type of automizer does not give a continuous spray, but spray fluid comes out only when it is pumped like a syringe. It has following parts:

(a) *Barrel:* It is hollow cylindrical and tin made barrel which measures 25 cm in length with a diameter of 4 cm. It is connected of the cap of tank on the lower side of the anterior end and this end of the tube is closed with a small hole in the middle, which is called nozzle. There is a tube (1/8" diameter) projecting from the tank and ending near just half of the nozzle and it travels upto the bottom of the tank and force the liquid in the spray form. The barrel has a small air hole (1/6" diameter) on its posterior end and it acts as an outlet and inlet for the air.

(b) *Barrel cap:* The posterior end of the barrel is also covered with a lid known as barrel cap which has a hole in its centre through which plunger rod moves.

(c) *Plunger rod:* It is made up of a wire (1/6" diameter) and provided with a wooden handle on the posterior end which helps in the operation of the plunger. On the anterior end of plunger, a washer and a nut are attached and in between, there is one leather valve, which closely fits within the walls of the barrel.

(d) *Tank:* It is rounded in shape and have the capacity of half pound. The tank is furnished with a cap known as filler cap.

Working: It operates almost in the same fashion as the cycle pump. First of all the liquid to be sprayed is filled in the tank. In the upstroke, when the plunger is pulled back, the air comes in the barrel through the nozzle and when plunger is pressed forward the pressure is exerted on the air. Since the leather valve is directed forward and is closely fitted within the barrel, the half portion of the nozzle is surrounded by the delivery tube. Therefore, the air of the delivery tube also goes out with the air of barrel and thus resulting in vacuum in the delivery tube. This is occupied by the spray liquid which in turn is thrown out in the form of a fine mist.

Hand Compression Sprayer

This type of sprayer is generally used to spray the field crops and trees. It is operated by a single man and it functions like a stove or petromax. This machine is available in two, three and four gallons capacity. By and lage the sprayer has the following important components:

(a) *Pump:* This pump operates exactly on the same principles as a football pump. It is fitted in between the upper end of the tank and hangs inside it. The pump has handle, plunger and plunger spring which are visible outside of the tank, where as barrel, leather valve and ball valve are found inside the tank. The function of the ball valve is to check the incoming of the air which is filled in the tank back into the pump.

(b) *Tank:* It has the cylindrical shape with the upper end convex and is made of brass or copper. The pump is fitted in the middle of the tank and the spray liquid is filled through a small opening having a diameter of 1 inch and covered by a filter cap which is situated on the upper end of the tank. The pressure gauge is attached just close to it which is meant to determine the pressure of air inside the tank. The bottom of the tank end is connected with other tube. The former tube is called delivery tube, this tube runs parallel to the bottom where as the latter one is a lance. In order to check the flow of the spray liquid by the operator, there exists one cut off cock between the joint of tubes. Besides, there is one stand that keep the tank in a right position and it is provided with a waist-plate. On the upper side of the tank, there is one circular plate around its body which is fitted with buckets. Each bucket is attached with the strap and at the distal end of each strap the hooks are provided. During spraying operation these hooks are connected with the waist plate.

Working: Before spraying operation, the liquid is filled in the tank preferable upto 3/4th of the total capacity of the tank. Thereafter the pressure is developed by pumping to the level of 55-60 psi (per square inch). Following this, sprayer is fitted on the back of the operator with the help of buckets, straps, hooks and wast plate. Then the cut-off cock is opened which is just near the bottom of the tank to fill the spray liquid in the form the spray of the liquid, the other cut-off cock which is fitted at the base of the extension rod is opened and thus the spray liquid is thrown out through nozzle in the form of a fine mist by the force of air pressure. In this type of machine the spraying is continued as long it is desired.

Paddle Pump Sprayer

This type of sprayer is more suitable to the field crops and particularly the trees which are upto the height of 20 feet.

An iron stand is provided as base to this sprayer. This machine is the assembly of a plunger, a stand suction hose, delivery hose and as extension rod having a spray nozzle at its free end. Since this is a foot operated sprayer, a paddle is attached to the plunger rod to make its up and down movement which ultimately operates the sprayer.

Working: No tank is provided to this sprayer but a separate bucket containing spray liquid works as a tank. A suction hose which is generally 2 m long and having a strainer at its free end is dipped in the spray liquid when the pedal is pushed down with the foot. The plunger is pushed up in the body of pump thus the air is compressed and moves out of the delivery tubes. The vacuum thus created results in the filling of the suction tube with spray liquid. Continuous paddling is required which sucks the liquid mixture into the body of the sprayer and is released through the delivery tube to the lance and finally through the nozzle which can be adjusted for desirable mist. One sprayer with a nozzle is enough to spray one hectare of a medium sized crop in day.

Power Sprayers

The manually operated sprayers are used for small holdings (up to 4 hectares) and for large scale spraying, power sprayers are required. Broadly, they may be classified in to three groups:

(i) Engine mounted power sprayer.

(ii) Tractor mounted power sprayer.

(iii) Aeroplane sprayer.

On the basis of amount of liquid sprayed and concentration of insecticide, they may be classified as follows:

(i) Low concentration and high volume sprayer.

(ii) High concentration and low volume sprayer.

(iii) Mist blowers and fog or aerosols generator.

Low Concentration Sprayer: This is the simplest form of power sprayer in which power is generated by a petrol or diesel engine of one or more cylinders and may have the capacity of 10 to 270 litres per minute. There is provision for overflow with a mechanical arrangements for agitation of spray liquid in the tank. The air chamber has a sturdy structure and helps in maintaining a high pressure of 25 kg per sq. cm. The engines are generally of 1 to 20 H.P. depending on the pump's capacity and pressure range. They draw the power is taken off the tractor. In such type of sprayers the spray liquid is forced out through the nozzle in small droplets under pressure developed by the pump. The droplets are deposited on the surface sprayed by the sheer force of the spray. The delivery hose varies from 25 m to 103 m or more in length and is provided with a hand operated spray gun which has a control valve mechanism for adjusting the spray pattern. In these sprayers the concentration of the spray liquid is low, quantity of spray used per hectare is about, 1,120 litres and the particle size is big.

High Concentration Low Volume: In these sprayers a relatively small volume of liquid is automised by a large volume of air high velocity. The principle involved is the same as in the case of hand automizer. They produce a much greater degree of automization then the nozzles used in low concentration sprayers as the spray liquid is discharged at 7 kg per sq. cm. These sprayers work by compressed air or have a rotary fan and a propeller turbine for production a current of air blowing at very high velocity into which a small quantity of spray material is introduced by gravity or with the help of a high pressure pump. In this case only 10-110 litres of the spray liquid is necessary for one hectare. Special nozzles are required for these sprayers and generally the insecticides are used in E.C. form because wettable powders may clog the nozzles.

Mist Blowers: These sprayers work on the same principle as the low volume high concentration sprayers ; the only difference being that the droplets are broken into particles of small size ranging from 50 to 100 micron.

The fog or smoke generators have been developed to disperse the spray material in extremely fine droplets (1 to 50 microns in diameter). The setting of aerosol droplets is very slow and as such is very much affected by the prevailing wind currents. This type of equipment can be successfully used where the foliage is dense and where fog takes time to settle and does not drift with the wind. A fog machine consists of tank with or without the engine and thermal or mechanical arrangements to break up the liquid into the desired droplet size and to create a force for carrying the fog from the machine.

Dusting Machinery

Hand Duster: This works on the simple principle of put forth the dust of air in which the dust particles are suspending. This type of duster is provided with a glass bottle and a piston to push the air into the duster. The dust is filled in the glass tube which comes out from the delivery hole with the compressed air. The use of this duster is limited to the small scale research plots or kitchen gardens.

Hand Rotary Duster

The appearance of this duster is larger as is operated by fitting it to the left side of the operator.

It is basically made up of a blower, gear box and a hopper. These are of hand carried type of shoulder mounted type of dusters.

Hopper: This is a drum shaped container which has the capacity generally of 5 kg. The hopper should not be filled more than three fourth of its capacity. A brast is attached on the back side of the hopper. Two banded iron bars known as agitator are connected with a crank shaft which passes through the centre of the hopper. These agitators move the dust continuously inside the hopper on the fan side which propels the dust into the blower through the exit hole situated in the hopper. To regulate the discharge of dust a feed regulator is provided on the outer side of this exit hole.

Gear Box: There is a box like structure attached to the blower on the right hand side of the drum and consist of four gears. Crank shaft is the main attachment to which the largest gear A is fitted in the centre of the box. This gear rotates a small double gear B with its grooved connections. Similarly, gear B rotates another double gear C and finally to smallest gear D which is attached to the fan shaft. This gear system helps in causing many rounds of gear D for one round of gear A and thus fan moves very fast. To operate the main gear one banded iron rod is linked with the crank shaft and called as crank.

Blower: This is a closed round shaped structure which opens to the opposite side of the drum through a small projected opening generally 1" in diameter. A fan provided with five blades arranged longitudinally is fitted in the centre of the blower. Rotation of this fan throws the air outside to create vacuum in the blower which is further occupied by the dust from the hopper and air from wire netting provided on the left side of the blower. The projected end of the blower is connected with a long tube known as lance which is fitted with a triangular reflector at its distal end.

Working: Dust should not be filled more than the 3/4 of the capacity of the hopper and this mist be closed tightly with its lid. The duster is tied to the left side shoulder of the operator with the help of straps provided with the duster keeping the breast plates on the chest. Lance is attached to the blower properly and reflector is also adjusted according to the need while operating the duster. Operator should move slowly forward in the direction of the wind rotation the handle continuously to ensure the uniform dust coverage of the crop.

There are other types of dusters also namely plunger type, blower type, knapsack type, fan type and power dusters, but they are not very common in use.

Power Dusters

These are operated by means of an engine or a tractor power take off device. A power duster comprises a hopper containing one or more agitators, a regulator for controlling the discharge of dust, a fan or blower producing an air blast for drawing in the dust from the hopper and blowing it out and flexible pipes with nozzles. The dust regulation device adjusts the dust flow at desired rate ranging from 1 to 10 kg per minute. The machines have arrangements for mixing a blast of air current with the powder. The mixed stream blowing out at velocity of about 320 km per hour.

The fans provided are mostly of centrifugal type. The fan case provided with one outlet or a series of four to eight or more outlets. These outlets are connected with flexible pipes to nozzle arranged on a boom in the rear of the tractor for carrying the dust. A power duster with one outlet can dust 12 to 16 hectares per day, depending upon the crop height and dosage.

Spraying Types: Based on the volume of spray mixture applied per unit area (at the rate of lit/hect) through different types of sprayer classified as high volume 750-1000 to be applied through manually operated sprayer, low volume 650-150, to be applied through motorised sprayer and less then 12 lits, to be applied through air craft or specially designed motorised/ battery operated equipments.

Dusting Versus Spraying

1. Dust can be used where water is limited or scarce.
2. Dusting is a quicker process as compared to spraying.
3. Sprays are more effective then dusts having the same amount of toxic chemical.
4. Sprays are not affected so much by wind as the dusts.
5. Excess of spray applied simply runs off, keeping safety value against excessive deposits.
6. The cost of dusts is comparatively low than the sprays.
7. The equipments required for spraying are comparatively costlier and heavier than the dusters.
8. In the case of dusting, uniform deposition is not possible, while in case of spraying it may be obtained upto a great extent.
9. Dusting is preferred in morning hours, while the spraying during evening hours.

OTHER MODES

Fumigators

These are generally used to kill insects or other pests by generating the gas through liquid or solid substances. There are different kinds of fumigators, but the commonly used one is ordinary fumigator cyno-gas pump which is described below:

Cyno-gas Pump

This fumigator is used for buffing the calcium cyanide powder or the cyno-gas. It has the following important parts:

Pump Cylinder: It is made up of brass tube which is 18" long and 3" in diameter. The upper end of this tube is closed by a pump cap and its lower end opens into a glass bottle through a small nozzle. This glass bottle is filled with the powder. In between the pump cylinder and glass bottle there is a small valve which allows the air to go into the bottle but checks its entry into the pump cylinder.

On the base of this pump cylinder there is one delivery tube which has got its connection with bottle. The air mixed with powder comes through this tube. Besides, one foot rest is also provided on the lower side of the pump cylinder.

A delivery hose made of plastic is connected with the delivery tube. In the posterior end of the delivery hose one thin brass rod is connected which measures one foot in length. The rod is known as lance.

Plunger Rod: The upper end of the plunger is connected with a wooden handle, while the other end has the leather valve which is fitted with washers etc.

Working: Cyno-gas pump is generally utilized in fields and godowns to kill rodents, ants and termites. First of all the bottle is filled up with the powder upto 3/4th of its capacity and there after the lance is inserted after slightly digging the ground of locating the burrows where the rats are present. Following this, the pumping is done and as a result the powder goes inside the burrows. Generally, two ounces of the poison is pumped in each burrow. When the operation is completed the holes should be closed with wet soil or by any other suitable means so that the rats may not come out.

Care and Maintenance of Plant Protection Equipments: The following measures may be observed to keep the machines in good working order:

1. Clean the machines properly and store them in dry place when not in use.
2. Flush the equipment with clean water to wash inside parts of containers, tubes and nozzles to keep them free from chemical.
3. The hoses and nozzles of the sprayers should be thoroughly washed with clean water after use and

then kept dry. Similarly, the dust from the duster should be taken out and cleaned with a cloth.

4. The moving parts should be greased and oiled regularly.
5. Keep the machines in working order by following a regular maintenance programme.
6. The oil should be kept at the required level in the power equipment.
7. Rubber hose should not be bent at angle when in use and should be removed and kept rolled when not in use.
8. Always stock spare parts, washers, nozzles etc. for replacements when the old parts get worn out.
9. Inspect pump valve seat, plunger cups and packing atleast once in a year to check leakage from these parts.
10. The spray liquid should be strained so as to avoid clogging of the nozzle and the delivery tube.
11. Test pressure regulator and pressure gauge periodically for efficient functioning of the sprayers.
12. Change oil regularly in the crank case to prevent corrosion of the bearings, gears and shafts.

Aerial Spraying

Aircrafts as a means of dispersing agricultural chemicals are now becoming popular as ex-military aircrafts, easily to adopt for agricultural work and ex-airforce pilots, desirous to continue flying, are available. This is supported by the availability of such pesticide which can be sprayed without dilution, making the spraying application economic. An aeroplane was first used in pest control in 1920 and now more than 40 countries are using aircrafts for this purpose. Aircrafts of the power ranging from 90-500 H.P. are used.

Spraying pattern varies with the nature of the chemical and compound-used, therefore the personal dealing with the spraying must be familiar with the term "high volume", low volume and "U.L.V. "etc. United states Department of Agriculture (USDA) has defined these terms as follows:

Full Coverage Spray or High Volume Spray: When used in labelling for agricultural chemicals, the term full coverage spray will apply when the total volume of spray to be applied will thoroughly cover the crop being treated to the point of run off or drip. High volume spray is applicable to low solubility or dispersibility of chemicals or highly soluble product. There is no danger of phytotoxicity but the availability of more labour and supply of water become a problem. Generally the high volume spray techniques require more than 400 litres of spray liquid/hectare.

Low Volume Sprays: When used in labelling for concentrated or technical agricultural chemicals the term low volume or concentrate spray will apply when the total volume of spray to be applied is adequate to uniformly cover the being treated, but not as a full coverage treatment to the point of run off. Low volume spray techniques are employed where spray liquid volume ranges in between 5 and 400 lit/hect.

Ultra Low Volume Sprayes: When used in labelling for concentrated or technical chemicals, the *term ultra* low volume will apply when the total volume of spray to be applied per acre is 1/2 gallon or less and is to be applied "undiluted". In all the ULV techniques less than 5 lit/hect. spray volume is applied. The various factors decide the choice of best suited techniques such as types of vegetation to be treated, kind of pests involved, availability of water, safety and experience of the operator.

ULV applications have been made generally by the aircraft equipped with a simple modification of the conventional boom and nozzle. For aerial spraying of the insecticides, both fexed-using aircrafts as well as helicopters are used. Correct nozzle tips and rotary automizers, their proper placement on the boom and maintenance of desired flying altitudes are the essential points in ULV application. The volume of insecticide used in ULV technique is extremely low and even a slight error in the above factors may result in inadequate or faulty coverage. At the normal operating speed of aeroplane flying at the height of five to seven meters a pattern of droplets 60 to ISO in diameter is laid down at densities of 15 to 20 meters, while crosswing of 8-11 kph are permissible. ULV application has proved its benefit for the treatment of large agricultural crops. ULV application is more faster thus considerably economic also. Another significant advantage of ULV application is that the biological performance in most cases is expanded for longer period. These two advantages have been achieved with ULV application of malathion. On the other hand all the insecticides are not suited for ULV application either due to their physical nature, high toxicity to man and animals or for lack of offering good biological performance. Some important desirable characteristics of an ULV product are: 1. It should have a very low volatility. 2. It should have a low viscosity. 3. Solvents (if used any) should have good solubility. 4. Solvent should be non-phytotoxic. 5. Solvent should be compatible with the pesticides. 6. The product should be safe for human being and animals.

On account of ideal characteristics for ULV application, malathion ULV concentrate is most widely used insecticide in various public health and agricultural programmes throughout the world. In India also, the usage of malathion ULV concentrate is well accepted. The recommendation of malathion ULV concentrate based on trials conducted in India; are given in the following table:

Crop	***Insect-Pests***	***Rate of Application***
Cotton	Jassids, thrips, whitefly, green weevil	360—480 ml/acre.
Paddy	Hispa, jassids, fulgorids	480 ml/acre.
Wheat	Flea beetle, pyrilla	480 ml/acre.
Sugar-cane	Pyrilla	480 ml/acre.
Vegetables	Thrips, Jassids, mealy bug, white fly epilachna beetle	360—480 ml/acre.
Jowar and other cereals	Deccan wing less grass-hopper, Ear head midge	480 ml/acre.

2

Destructing Insects

Crop Destroyers

Kharif Grasshopper

Local Name: Phangi, Kata, Tota, Pharka, Ankhfura and Tidda.

Scientific Name

(i) *Hieroglyphus nigrorepletus Bolivier*

(ii) *Hieroglyphus banian Fabr.*

Order : Orthoptera

Family : Acrididae

Host Plants: Paddy, Jowar, Bajra, Maize, Sugarcane, Smaller millets, Sunhemp, Arhar and Grasses.

Distribution: Besides India, it is found in Mayanmar, Sri Lanka, Malaya and Hawaii islands. In India it is distributed in all paddy growing areas.

Nature of Damage: Pest appears in the field on the, onset of monsoon. Initially it feeds on grasses and as soon as Kharif crops are grown, it starts feeding on them. The damage is caused by both adults and nymphs having chewing and cutting type of mouth parts. In comparison to adults, nymphs cause more damage because nymphal period is longer varying from July to September and secondly they eat more to fulfill the various activities (growth etc.) of the life while adults feed only for their survival.

They feed on leaves and shoots and in case of severe infestation they completely defoliate the plants leaving only the mid-ribs. Due to which the growth of plants badly affected and remains stunted resulting in poor yield. At the time of milky stage in rice, they also feed on ears thus causing heavy loss. The severe damage is usually done during August and September. Though it is a sporadic but serious pest in most of the eastern Uttar Pradesh. Few years back it appeared in epidemic form in Rajasthan and completely destroyed the crops. It has been estimated that an individual grasshopper during its life time eats about 42 grams of green maize leaves.

Life Cycle: Three developmental stages, *viz.*, egg- nymph and adult, are found in its life cycle.

Egg: The eggs are deposited by the female in masses of 30-40 at a depth of 5 to 7.5 cm in the soil along the side of field bunds from October to December. When the female is ready to lay eggs it trails its abdomen on the ground for searching of suitable place. After finding a suitable place the female drills the surface of the soil with its ovipositor and forms a hole for egg laying. When hole is ready it secrets a fluid from gonopore and starts egg laying one by one in it. The abdomen is withdrawn as the hole gets filled with eggs. This hole is called as egg chamber.

The female again secretes a fluid on the eggs and then filled up the egg chamber with the loose soil by its hind legs. The foamy material exuded by the female serves for sticking the eggs together and for cementing the soil particles which constitute a pod-like structure with eggs within it. These egg masses are known as egg pod which remain in the soil till the onset of rains in June-July of the following year. Generally single female lays 15 to 20 egg masses in her life span. An individual egg is light yellow in colour and resembles to paddy grain in shape. At the time of hatching colour changes in to brown. The egg period is of about 9 months.

Nymph: The newly hatched tiny numphs are light yellow in colour measuring 6 mm long and devoid of wings. They start feeding on grasses and can be seen jumping from one place to another. The nymphal period is of 70 to 80 days and during this period they cast their skins 5 or 6 times at an interval of 10 to 15 days. After second moult the colour changes to light green and wing pads are visible.

Adult: The adult hoppers are yellowish-green in colour. *H. nigrorepletus* is larger than *H. baninan;* the male and female of the former are 37 mm and 75 mm long while of the later 25 mm and 37 mm long. In case of *H. banian* there are only 2-3 black markings but in *H. nigrorepletus* there is a reticulum of black lines and possesses U shaped black spot on the lateral sides of pronotum.

The adults are seen in the last week of September or early October. They copulate and may be seen for hours in same position. The males die after copulation while females after egg laying. Only one generation is found in a year. After becoming adults the insects take 7 to 20 days to attain sexual maturity and after the first mating they require a period of about 7 days to lay the first batch of eggs. The life of female adult lasts for about two months.

Control

1. Plough the affected fields and the bunds to a depth of 10-15 cm before the monsoon begins, it will destroy the egg masses.
2. The hoppers and adults can be collected and destroyed.
3. At the time of on set of monsoon they may be attracted by light-traps and destroyed.

4. The dusting of following insecticides has been found effective in minimising the population of pest:
   (a) Methyl parathion 2% (3) 30 kg/hect.
   (b) Fenitrothion 5% @ 20 kg/hect.
   (c) Malathion 5% @ 30 kg/hect.
5. The spraying of following insecticides has also been found quite effective:
   (a) D D V P 76 EC-200 ml/hect.
   (b) Malathion 50 EC-2.5 lit/hect.
6. Poison bait with following ingredients may also be prepared and spread in the field to control the hoppers:
   (a) Bran-10 kg.
   (b) Sodium arsenate- 1/2 kg.
   (c) Mollases-2.5 lit.
   (d) Water-10 lit.

Natural Enemies: The following are the egg parasites of this pest:

(a) *Cacallus sp.*

(b) *Barycomus sp.*

(c) *Scelio sp.*

Some birds like crow and kite eat the nymphs and adults while others like wood-pecker and blue-jay eat the eggs of the hoppers.

Gundhi Bug

Local Name: Rainya, Chusia, Gundhi

Scientific Name: *Leptocorisa varicornis* Fabr.

L. costalis

L. acuta

Order : Hemiptera

Sub-order : Heteroptera

Family : Coreidae

Host Plants: Paddy, jowar, bajra, maize, smaller millets and many grasses.

Distribution: It is found in most of the rice-growing areas of India and its distribution extends over the whole of South-Asia from China and Japan in the north to Sri Lanka in the south. It is also found in the Philippine islands and Australia. In India its attack is specially severe in Uttar Pradesh and Bihar.

Nature of Damage: It is the most serious bug pest of the paddy crop. Both nymphs and adults suck the juice initially from the leaves and tender shoots and later from the ears. Although it feeds on a variety of plants but creates serious problem when grain-formation is in progress in the paddy. It sucks the milky juice, leaving the white chaffy husk. A hole is left on the grain where bug punctures with its proboscis and around which a brownish spot develops due to fungus attack. Its attack can easily be recognised by unpleasant odour which comes from the field where it is present and the due to which name 'gundhi' has been given. The insect is very common i.e. paddy field from middle if August to middle November and causes 20 to 30% damage of crop. In case of severe infestation, practically all the grains of all the earheads become chaffy. During 1958 it appeared in epidemic from in Assam. West Bengal, Bihar, Madhya Pradesh and U.P. and control operations where under taken in 4,84,168 acres of paddy fields in U.P.

Life Cycle: Three developmental stages namely egg, nymph and adult are found in its life cycle which are described as under:

Egg: The eggs are laid by the female bug in the month of July and August. They are deposited in three stripes on the lower side of the leaves. Each stripe contains 10 to 20 eggs and about 40 eggs are laid at the place. The eggs are blackish-brown small bead like in structure. Depending on the prevailing temperature, the eggs hatch in about 4-7 days. At the time of hatching their colour changes into black.

Nymph: The slender greenish nymphs hatch out from the eggs, start immediately sucking the plant juice. At the time of hatching it is about 1.8 m long which grows up to 2.0 mm within 6 hours. There are five moults after which the nymph becomes adult with wings. The wings-pad may seen in the IIIrd nymphal instar. The nymphal period is of 14 to 20 days. The full grown nymph measures 15 mm long.

Adult: The adult bug is about 15-16 mm long slender and greenish-yellow in colour. Head is small and provided with 4 segmented rostrum. The antennae are larger than the body and red in colour. Meta-thorax provided with glands emitting unpleasant odour. The fore wings are hemelytrat.

The life cycle is completed in 21 to 31 days and several overlapping generations are found but on paddy 4 to 5 generations are completed. The bugs copulate after 12-14 days of emergence and female survive up to 55 days while male only for 33 days. The pest passes its winter in adult stage and migrate to grasses and other plants to continue it's life cycle.

Control

1. All weeds and grasses growing around the paddy field should be destroyed.
2. The nymphs and adults may be attracted in light-traps.
3. Resistant varieties should be preferred in infested areas, *e.g.*, Sathia in U.P., Sona in Bihar and Mundga kuttin in Chennai.
4. The dusting with any one of the following insecticides has been found quite effective:
   (a) Methyl pasathion 2% dust @ 25 kg/hect.
   (b) Malathion 5% dust @ 30 kg/hect.
   (c) Thiodan 4% dust @ 20 kg/hect.
   (d) Carbaryl 10% dust @ 20 kg/hect.

Natural Enemies: The important natural enemies of the pest are given as follows:

(a) Mantis-feeds on nymph and adults.

(b) The following coccinellids also feed on nymphs and adults:

(i) Coccinella undecipunctata

(ii) Adalia sp.

(iii) Ommatius sp.

(c) Pentatomid and reduvid bugs also prey on them

Fruit Fly

Common Name: Son Makhi, Fal mashi, Fal makhi etc.

Scientific Name: *Dacus (Strutmeta) cucurbitae Coq.*

Order : Diptera

Family : Trypetidae

Host Plants: Cucurbitaceous plants such as gourd, pumpkin, cucumber, Kerala, taroi, melon, khera, tinda and some times tomato, papaya, brinjal and guava.

Distribution: The fruit flies are widely distributed through out the world. It was first seen in Hawaii Island during 1898 and by 1913 spread all over the world.

Nature of Damage: The damage is caused by the maggots of the fly and considered to be a group of very harmful insects which cause serious damage to fruits and vegetables. Generally, fruits are attacked in the early immature soft stage with the result that; they do not develop, drop down and rot. Those which do not drop and are picked up, found unfit for human consumption as they have a number of abominable white maggots in the pulp. The humid-hot weather is most suitable for its attack and generally 50 to 60 percent loss is caused to vegetable crops.

Life Cycle: There are four developmental stages in its life cycle which are described as follows:

Egg: The female fly lays eggs from February to November. The pre-oviposition period is 14 days in August and 9 to 30 days during other months depending upon the environmental conditions. The fly lays small, shiny whitish cigar shaped eggs in small clusters just under the skin of the fruits with the help of her ovipositor and measuring 0.6 mm long. Each cluster may have 4 to 10 eggs and they are laid in many times at an interval of 2 to 5 days. Female lays about 300 eggs in her life span. The hatching periods of eggs varied from 2 to 9 days.

Larvae (Maggots): The maggots on hatching feed on the fruit pulp by burrowing into it. They are white in colour and measuring 1 mm long. Their body is divided into 11 segments and one end of the body is wider than the other. They are devoid of legs. The maggots spend their whole larval period in side the fruit and by this time, the attacked fruits are generally rot and the larvae leave the rotten fruit and enter the soil for pupation. The larval period takes 7 to 8 days but may be 3 days in summer and 20 to 21 days in winter. Full grown maggot is 19 mm long and 2 mm wide.

Pupa: Pupation takes place in the soil few cm below the surface. The pupa is flat, oval pointed towards anterior and flat to posterior end. It is 11 segmented and 5.5 mm long with light yellow in colour. The pupal period generally lasts 6-9 days which may extend 20-21 days in November and 28-30 days during December.

Adult: The fly is reddish-brown in colour with black and white spots. The wings have brown bands and spots at the apex. It measures about 7 mm in length and 3 mm in breadth with a wing expansion of about 14 mm. The head is like a top provided with aristate antennae. There are 3 bright yellow stripes on the meta-thorax.

The life cycle is completed in 15-20 days during summer and rainy season which extends to 60 days in winter. There are many overlapping generations in a year. The adults are generally long lived, male survives about 56 days while female for about 66 days. They remain in hibernation during November to January.

Control

1. The ploughing of fields two to three times reduces the pest population by exposing the pupae to enemies.
2. Collection and destruction of infested fruits along with the larval stage of the fruit-fly.
3. Baging of fruits protects the fly from egg-laying.
4. Dusting the plants with tobacco dust or ash mixed with kerosine oil to repel the flies.
5. Male flies may be attracted in citrenela oil and destroyed.
6. The adult flies may also be controlled with the spray of following solution:
   (a) Tartar acid-1 part
   (b) Water-320 parts
   (c) Mollases- 24 parts
7. Use bait-traps prepared with the following. This can be put in small tin containers which are hung from sticks at various places in the field:
   (a) Protein hydrolysate-100 g.
   (b) Water-10-20 lit.
   (c) Malathion 25% W.P.-200 g.
8. Some chemosterilants, *viz.*, tepa, metepa and apholate have been found effective in sterilizing the fly which will prevent the multiplication of pest.
9. The following insecticides may be sprayed regularly during the active period of the fly.
   (a) Malathion 50 EC/W.P.-2.0 lit or 2.0 kg/hect.
   (b) Formothion 25 EC-0.650 lit/hect.
   (c) Carbaryl 50 W.P.-3.0 kg/hect.
   (d) DDVP 76 EC-0.250 lit/hect.

Natural Enemies: The following are the pupal parasite of the pest:

(a) *Opius fletcheri* Silv.

(b) *O. incisus* Silv.

(c) *O. compensans* Silv.

Red Pumpkin Beetle

Local Name: Torai ka kira, Lal kira

Scientific Name: Three species of the insects are common:

(a) *Raphidopalpa foveicollis Lucas* — *Red coloured*

(b) *R. intermedia Jaobx.* — *Black round*

(c) *R. cincta Fabr.* — *Brown coloured*

Order : Coleoptera

Family : Chrysomelidae

Host Plants: All the cucurbitaceous plants such as melon, cucumber, gourd, torai, tinda, kheera, and kakari, etc.

Distribution: Besides India; it is distributed Ceylon, Pakistan, Mayanmar, Sudan, Italy and Australia. In India, it is most common in U.P., Punjab. M.P. Maharashtra, Bihar and Rajasthan.

Nature of Damage: The damage is caused by both grubs and adults of the insect. The adult beetles feed on the leaves, flower buds and flowers, there by causing severe injury specially to young plants. The damage done to young seedlings is often so severe that the crop has to be resown because all the leaves of young plants are eaten up. In older vines the beetles eat and make holes into the leaves. The grub bores in to the root and stem of the plant and destroy the whole plant when present in large number. The attacked plants wither and die. They also attack the fruits near the portion resting on the ground and become unmarketable.

Life Cycle: Four developmental stages namely egg, larva (grub), pupa and adult are found as described below:

Egg: Hibernating beetles come out in the beginning of March and copulate after 7 to 8 days of emergence. Pre-oviposition period is of 2-7 days. The females deposit orange-coloured eggs singly or in small clusters on moist soil near the base of the seedlings. A single female is able to lay 60-300 eggs in her life span. The incubation period of the eggs varies from 5-15 days depending on environmental temperature and humidity.

Grub: The grubs soon after hatching are quite active and begin to feed on the roots and stems which readily bore in to. They are creamy-yellow in colour with brown head and prothorax. The last segment is dark coloured and is provided with a medium cylindrical process which is used by the grub in walking. At the end of every instar, the larvae enter the soil to moult and come out there after to feed again. There are four larval instars before it becomes full grown in 13-25 days. The full grown grub is about 15 mm long.

Pupa: At the end of the last instar when the larva is full grown, it tunnels in to the soil to depths varying from 1 cm to 2.5 cm and there it prepares an oval chamber for pupation. The pupa is reddish-brown in colour and measuring 7.5 x 3.5 mm. The pupal period varies from 7 to 14 days.

Adult: The adult beetle measures about 7 mm in length and 4 mm in breadth. The upper surface of the body is bright orange-red while lower surface of the abdomen is entirely black and covered with short whitish soft hairs. The posterior portion of the head is covered with pro-thorax.

The adult beetle soon after emergence begins to feed and starts egg laying after a week. The life cycle is completed in 32-61 days and there are 5 to 8 generations in a year. During winter the beetles remain in hibernation hiding in various niches from the severity of the cold. As soon as it begins to warm up, the over wintering population becomes active and comes out from its hiding places.

Control

1. Plough the fields after harvest to destroy the pupae and larvae therein.

2. Early sowing of cucurbitaceous plants, *i.e.*, in November protects the crop from the beetles appearing after hibernation as the plants are well established in the mean time.
3. Dusting the crop with kerosinized ash, will repel the beetles.
4. Interculturing around the roots of vines in the month of March-April minimises the attack of grubs.
5. The vines may be dusted with any one of the following insecticides:
   (a) Malathion 5% dust @ 15-20 kg/hect.
   (b) Carbaryl 5% dust @ 20 kg/hect.
   (c) Endosulfan 4% dust @ 15-20 kg/hect.
6. Sprayings with following insecticides have also been found effective:
   (a) Malathion 50 EC/Wp- 1.25 lit or 1.25 kg/hect.
   (b) Formothion 25 EC-0.650 lit/hect.
   (c) D D V P 76 EC-0.250 lit/hect.

Natural Enemies: Some birds like crow, myna and blue jay devour the larvae and adults of the pest. Preying-mantis has also been observed preying on adults.

Epilachna Beetle

Scientific Name: *Epilachna vigintioctopunctata* (Fabr.)

E. dodicastignia (Mulsant)

Order : Coleoptera

Family : Coccinellidae

Host Plants: Potato, brinjal, tomato, bitter gourd and other cucurbitaceous and solanaceous plants.

Distribution: Through out India.

Nature of Damage: The damage is caused by both adults and grubs which feed on the upper surface of the leaves. They skeletonize the leaves which present a lace-like appearance as the green matter in between the veins is eaten away leaving the skeleton of anastomosing veins. They eat up regular areas of the leaf tissue, leaving parallel bands of uneaten tissue in between. The leaves turn brown, dry up and fall off, completely detoliating the plants.

Life History: The four developmental stages namely egg, larva, pupa and adult is found in its life cycle which have been described below:

Egg: The eggs are generally laid on the under surface of leaves in a prominent yellow batches of 5-40 eggs. They are cigar shaped and glued to the leaf surface in a vertical position. A single female can lay upto 400 eggs in her life span. The hatching period of eggs varies from 3-4 days which may extend upto 9 days in winter.

Larva: The newly hatched grub is stout and yellow in colour, bears spines all over the body. They feed on the lower opidermis of the leaves. The full grown grub measures 8 mm in length and 4 mm in breadth, the body being brood in front and narrow behind and covered with spiny structures all over. The larval period varies from 12-18 days during which there are three larval instars.

Pupa: The pupation takes place on the leaf surface or around the base of the stem. At that time the full grown larva attaches the last segment of its abdomen to the leaf surface by means of a sticky secretion and the pupa is formed within the last larval skin which splits on the dorsal side. The pupa is yellowish-orange in colour with brownish white markings on the upper surface. The pupal period varies from 3-6 days.

Adult: The adult is yellowish-brown in colour convex, oval shaped and bears small, round black spots on the elytra. The *E. vigintioc-topunctata* bears 14 such spots on each elytra while *E. dodicastigma* possesses 6 spots on each elytra. Adult longevity varies from less than a month to more than two months. The beetle passes winter in hibernating stage in the heaps of dry plants or cracks and crevices in the soil.

The total life cycle is completed in 19 to 30 days and there are 7-8 generations in a year, while in hills only one generation is completed. The pest remains active from March to October and during hot and dry months, the number decline very much, but the population builds up again in August.

Control

1. Collect and destroy the egg masses and other stages of pest during early stage of infestation.
2. Dust the crop with any one of the following insecticides per hectare:
   (i) Malathion 5% dust - 25 kg/hect.
   (ii) Carbaryl 5% dust - 30 kg/hect.
   (iii) Endosulfan 2% dust - 25 kg/hect.
3. Spray the crop with any one of the following insecticides:
   (i) Endosulfan - 0.05%
   (ii) Methyl parathion - 0.02%
   (iii) Malathion - 0.05%

Brinjal Shoot and Fruit Borer

Scientific Name: *Leucinodes orbonalis* Guen.

Order : Lepidoptera

Family : Pyralidae

Food Plants: Brinjal and other solanaceous plants.

Distribution: The pest is widely distributed in Malaysia, Sri Lanka, India, Pakistan, Germany and East Africa.

Nature of Damage: The damage is caused by the caterpillar which attacks the top shoots of young plants as well as fruits. The larvae bore into the shoot and eat the internal tissues causing the attacked portion to droop and wither away. When the terminal shoots are attacked the growing points are killed. In young fruits, the newly hatched larvae bore into the fruit and the entry hole is too small to be easily noticeable. The large holes usually seen on the fruits are the exit holes of the caterpillars. The infested fruits become unfit for human consumption. Damage to the fruits, particulary in the autmn˙ is very severe and the whole crop may be destroyed by the borers. A single caterpiller may destroy 4-6 fruits.

Life History: The four developmental stage *viz.*, egg, larva, pupa and adult ore found in its life cycle which have been described below:

Egg: The female moth lays eggs singly or in batches of 2-4 on the underside of leaves, on green stems, flower buds or the caly of fruits. The eggs are creamy-white in colour and elongate. Each female generally lays more than 150 eggs. The hatching period of eggs varies from 3-6 days.

Larva: The young caterpillars bore into tender shoots near the growing pointer, into the flower buds or into the fruits. The young caterpillar is creamy white in colour which becomes pale white with violet mashings arranged in linear lines all over the body when full grown. The full grown caterpillar measures about 18-23 mm in length. The larvae grow through 5 stages and larval period losts for 14-20 days.

Pupa: The full grown caterpillars comes out of the fruits or shoots and pupate amongst fallen leaves in a boat shaped silken cocoon. The pupal period lasts for 7 to 11 days after which the moths emerge.

Adult: The moths are medium sized insects with white colour but there are pale brown or black spots on the dorsum of thorax and abdomen. Its wings are white with a pinkish or bluish tinge and are ringed with small hair along the apical and anal margins. The moth measures about 20-22 mm across the wings. The fore wings are ornamented with a number of black, pale and light brown spots.

The total life cycle is completed in 25-39 days and there are 5 overlapping generations in a year. The caterpillars hibernate in the winter and pupate early in the spring. The moths appear in March-April to continue the next generation.

Control

1. Collect and destroy the drooping shoots and affected fruits which contain caterpillars inside.
2. Spray the crop with any one of the following insecticides:
   (i) Endosulfan - 0.07%
   (ii) Carbaryl - 0.15%
   (iii) Malathion - 0.05%
   (iv) Fenvalerate - 0.003%

Tobacco Caterpillar

Scientific Name: *Spodoptera litura* Fabr.

Order : Lepidoptera

Family : Noctuidae

Host Plants: Tobacco, tomato, castor, mustard, cabbage, cauliflower, knolkhol, potato, gram, pea and chillies etc.

Distribution: It is found throughout the tropical and subtropical parts of the world. The pest is widely distributed in India.

Nature of Damage: The damage is caused by the caterpillar which feed on the leaves and fresh growth of the plants. They are voracious feeder of leaves and several of them when present on a plant cause defoliation. The pest in active in the night and feeds by biting large holes in the leaves. Some times whole crop is destroyed due to their attack particularly in tobacco nurseries.

Life History: The four developmental stages, *viz.*, egg, larva, pupa and adult are found in its life cycle which are described below.

Egg: The moths are active at night when they mate and female lays eggs in cluster on the under surface of the leaves. The eggs are covered with buff-coloured hairs. Each cluster may have 100-250 eggs. The hatching period of eggs varies from 4-5 days.

Larva: The newly hatched larvae feed gregariously for first few days and then disperse to feed individually. The full grown caterpillar is about 35-40 mm long and is velvety black with yellowish-green dorsal stripes and lateral white bands. They pass through 6 stages. The larval period lasts for 15-30 days.

Pupa: When the larva is full grown, it comes down off the plant and pupates in the soil. The pupa is reddish brown in colour and measures 20-22 mm in length. The pupal period lasts for 7-15 days.

Adult: The moth is a medium size, dark brown in colour with irregular whitish lines on the lore wings. The hind wings are white. They are about 22 mm long and measure 40 mm across the spread wings. After emergence they live for 7-10 days.

The life cycle is completed in 27 to 52 days and there are about 8 generations in a year. The pest breeds throughout the year, although its development is considerably reduced during the winter.

Control

1. During early stage, when they feed gregariously may be collected and destroyed.
2. Spraying the crop with anyone of the following insecticides:
   (i) Malathion – 0.05%
   (ii) Carbaryl – 0.15%
   (iii) DDVP – 0.03%
   (iv) Endosulfan – 0.07%

Cabbage Butterfly

Scientific Name: *Pieris brassicae* Linn.

Order : Lepidoptera

Family : Pieridae

Host Plants: Cabbage, cauliflower, knolkhol, turnip, radish and mustard etc.

Distribution: The cabbage butterfly is world-wide and is found wherever cruciferous vegetables are grown.

Nature of Damage: Damage is caused by the caterpillar which feeds on the leaves of the plant. The first instar caterpillars just scrap the leaf surface, whereas the subsequent instars eat up leaves from the margin inwards leaving intact the main viens. When the caterpillars are abundant they leave only the bare stalks of the leaves on the plants.

Life History: The four developmental stages, *viz.,* egg, larva, pupa and adult are found in its life cycle which are described below:

Egg: The female butterfly is vary active in the field and lays a large number of eggs on the leaves of food plant. On an average, 164 yellowish conical eggs in clusters 50-90 eggs are laid on the lower or upper surface of the leaves. The hatching period of eggs varies from 3-7 days in March-May and 11-17 days in November to February.

Larva: The small green caterpillars with black heads come out from the eggs and start feeding on the leaves. The caterpillars feed gregariously during the early instars and disperse later on. The larvae are pale-yellow, when young and become greenish yellow when full grown. The head is black and the dossum is marked with black spots. The full grown caterpillar measures about 40-50 mm in length. They moult four times and become full grown in 15-22 days during March-April and in 30-40 days during November to February.

Pupa: The full grown caterpillar pupates either on a leaf or on a tree or at any appropriate place with a girdle of silk round the middle to hold the chrysold in position. The pupal stage lasts 8 to 15 days in March-April and 20-28 days in November to February.

Adult: The adults are pale white butterflies and have a smoky shade on the dorsal side of the body. The wings are pale white, with a black patch on the apical angle of each fore wing and a black spot on the costal margin of each hind wing. The male is small than female. The female below larva measure 6.5 cm across the spread wings. They live for 2-12 days depending upon the season.

The pest appears on cruciferous vegetables in the beginning of October and remains active upto the end of April. It is not found in plains from May-September during these months it breads in the hills. The life cycle is completed in 27-87 days and there are 4 generations in a season (October to April).

Control

1. Collection and destruction of egg masses and larvae during early stage.
2. Spray the crop with any one of the following insecticide:
   (i) Malathion - 0.05%
   (ii) Endosulfan - 0.07%
   (iii) Carbaryl - 0.15%

Diamond Back Moth

Scientific Name: *Plutella xylostella* Linn.

Order : Lepidoptera

Family : Plutellidae

Host Plants: Cauliflower, cabbage and rape-seed etc.

Distribution: It is world-wide in distribution and found throughout Indian union.

Nature of Damage: The damage is caused by the caterpillars which is in the earlier stages, feed in mines on the lower side of the cabbage leaves and in the later stages on every position of the leaves. Central leaves of cabbage and cauliflower may be riddled and the vegetables rendered unfit for human consumption. They particularly feed in the heart or head of cabbage and cauliflower. It is most serious when appears in the early stages of the crops, *i.e.*, during August-September.

Life Cycle: The four developmental stages, *viz.*, egg, larva, pupa and adult are found which are described below:

Egg: The eggs are laid by the female moth on the under side of the leaves. They are usually laid singly or in batches of 2-40, which are yellowish in colour. The single female can lay about 356 eggs in her life span. The hatching period of eggs varies from 3-9 days.

Larva: The newly hatched larvae bore into the tissues from the underside of the leaves and feed in the tunnels. In the third instar, the caterpillars feed outside the tunnels and in fourth instar feed from the underside of the leaves, leaving intact a parchment-like transparent cuticular layer on the dorsal surface. The full grown larva measures about 8 mm in length and are pale-yellowish-green, with fine black hair scattered all over the body. The larval period lasts for 8-16 days.

Pupa: The full grown larva constructs a barrel-shaped silken cocoon, which is open at both ends and is attached to the leaf surface. The larva pupates inside the cocoon and comes out in 4-5 days as an adult.

Adult: The adult measures about 8-12 mm in length and is brown or grey in colour. They have conspicuous white spots on the fore wings, and appears like diamond patterns when the wings lie flat over the body. The moths may live for as long as 20 days.

The pest is active throughout the year. The life cycle is completed in 17-23 days and there are several generations in a year.

Control

The pest may be controlled by the sprayings of anyone of the following insecticides at weekly intervals:

(i) Malathion - 0.05%
(ii) Endosulfan - 0.07%
(iii) Fenitrothion - 0.05%
(iv) Carbaryl - 0.15%

Mustard Aphid

Local Name: Chenpa, Mahun, Mahon, Tela, Mawa, Moyala, Lahi.

Scientific Name: *Lipaphis erysimi* Kalt. (*Siphocoryne indobrassicae*)

Order : Hemiptera

Sub order : Homoptera

Family : Aphididae

Host Plants: Mustard, cauliflower, cabbage, knolkhol, radish and turnip etc.

Distribution: It India it is found in all the states specially in U.P., M.P., Punjab, Maharashtra, Bihar, Gujarat, Mysore and Rajasthan.

Nature of Damage: In appears in the last week of November or early December in U.P. and reaches to its peak in the month of February. Both nymphs and adults suck the sap from the tender portion of the plants. Hundreds and thousands of these may be seen on leaves or tender shoot. In mild case, the shoot wilts and in severe cases it completely gets killed. The honey-dew produced by such a large number of individuals covers practically the whole surface of the tender leaves and shoot and a kind of black mould develops which interferes the photosynthetic activities of plant. Honey-dew secreted by aphids is liked by a number of other insects specially by ants who for the sake of ensuring the supply of this wet liquid, protect the aphids and also provide them transport; this type of association is known as commensalism. The cloudy and humid atmosphere is most suitable for its multiplication and attack.

Life Cycle: A different type of life-cycle has been found in plain and hill regions. In plains, the insect does not lay eggs hence only nymphs and adults are found.

In Plains: They appear in November-December on mustard plants and among them females are more in number than males. The females are viviparous and they do not lay eggs but produce many nymphs. The nymphs are produced by following ways:

(a) After mating the males.

(b) Without mating the males but parthenogenetically.

The nymphs become adult within 3 to 7 days and start producing apterous offsprings. In this way they increase great in number and spread over a big area in a very short time. There are several generations during the cold season.

In Hills: In hills the life cycle of aphids is just like in European country. Aphids are remarkable on account of their peculiar mode of development and the polymorphism exhibited in different generations of the same species. The following types of individuals are present in the life cycle of migratory aphids:

Fundatrices: These are apterous, viviparous parthenogenetic females which emerge in spring from the over-winter and eggs. Only one generation is found of this group.

Fundatrigeniae: These are also apterous, parthenogenetic, viviparous females and are the progeny of fundatrices. They live on the primary host and complete 3 generations.

Migratory: These usually develop in the second, third or later generations of fundatrigeniae and consist of winged parthenogenetic, viviparous females. They develop on the primary host and subsequently fly to the secondary host.

Alienicolae: Parthenogenetic, viviparous females developing for the most part on the secondary host. They often markedly from the fundatrices and migrants. They are similar to fundatrigeniae but differ from them by living on secondary host while former on primary host. Many generations may be produced comprising both apterous and winged forms.

Sexuparae: Parthenogenetic, viviparous females which develop on the secondary host and migrate to the primary host at the end of summer.

Sexuales: These are the progeny of sexuparae produced on the primary host. They consist of sexually reproducing males and females. The females lay eggs on the primary host after copulating the males which passes winter as such and hatch out in following spring and fundatrices females are produced.

Adult: The aphids are small generally 2 mm in size, green sucking insects. They are provided with a pair of small tubular structure projecting out from dorsal surface of the posterior region of the body known as cornicles or siphons or honey tubes. Two pairs of transparent wings are found in which costa and sub-costa veins are absent.

Control

1. Early sowing or early maturing varieties should be grown to escape the damage.
2. Resistant varieties *viz.,* AGH-4 and T. 101 should be preferred to grow.
3. The crop may be dusted with any one of the following insecticides:
   (a) Malathion 5% dust @ 25 kg/hect.
   (b) Carbaryl 10% dust @ 25 kg/hect.
   (c) Fenetrothion 5% dust @ 20 kg/hect.
4. The spraying of any one of the following insecticides has been found effective to control the aphids:
   (a) Phosphomidon 100 EC @ 0.250 lit/hect.
   (b) Dimethoate 30 EC @ 1.00 lit/hect.

(c) Endosulfan 35 EC @ 1.25 lit/hect.

(d) Formothion 25 EC @ 0.650 lit/hect.

(e) Methyl demeton 25 EC @ 1.00 lit/hect.

(f) Malathion 50 EC @ 0.600 lit/hect.

(g) Chlorpyrifos 20 EC @ 0.600 lit/hect.

Natural Enemies: The following insects feed on aphids:

(a) *Coccinella septempunctata (Lady bird beetle)*

(b) *Chrysopa sp. (Aphid lion)*

(c) *Syrphid sp. (Maggots of syrphid)*

Mustard Saw-fly

Common Name: Sarson ki illi

Scientific Name: *Athalia proximo* Klug.

Order : Hymenoptera

Family : Tenthredinidae

Host Plants: Mustard, radish, turnip, cabbage, cauliflower and other cruciferous plants.

Distribution: The pest is distributed practically throughout India. Besides India, it is widely found in Indonesia, Formosa and Mayanmar.

Nature of Damage: In most parts of the country it is pest of cold weather and its activity confined generally to the period from October to March although in certain western regions it is reported to be prevalent round about August. The larval stage damages crop by voractiously feeding on the young crop. Its feeding activities are generally confined to mornings and evenings. As a result of feeding the leaves become full of holes and the plants either dry up or remain stunted. In case of severe infestation all the leaves are eaten and the plants die.

Life Cycle: Four developmental stages namely egg, larva (grub), pupa and adult are found in its life cycle.

Egg: The female lays eggs singly in to the tissues of the food plant. The egg laying organ (ovipositor) is highly specialized for silting open the margin of the leaves within which eggs are laid. Initially they are milky in colour which turn in to dark brown or black at the time of hatching. An egg is 2 mm long and many eggs may be laid in to one leaf. Each female is capable of laying usually 30-61 eggs which in some cases may reach up to 150 eggs. Hatching period of eggs varies from 4 tò 27 days depending upon the environmental conditions.

Larva (grub): The young grub is greenish grey in colour and its body surface is hairless measuring 2 mm long. It begins to feed on the margin of the leaf, and as it grow, its colour getting darker and darker. Full grown grub is 15 to 20 mm long having 3 pairs thoracic and 8 pairs abdominal legs and provided with 5 stripes on the back. The larval period lasts for 12 to 18 days.

Pupa: For pupation grub goes in to soil wherein it prepares cocoon made of silk buttressed by soil particles. The cocoon is 7-11 mm long and 4 to 6 mm wide. The pupal period is of about 10-12 days. During severe cold it hibernates in pupal stage.

Adult: The adult sawfly has a short thick-set body with a mixture of yellow brown markings on it and a dark reddish brown colour on the wings. It has two pairs of wings and good flier, its activities are diurnal. The fly is also capable to lay eggs parthenogenetically, but only males emerge out from these eggs.

Control

1. Hand picking of grubs early in the morning is found useful to check the population of the pest.
2. Dusting the crop with any one of the following insecticides has been found effective:
   (a) Malathion 5% dust @ 25 kg/hect.
   (b) Thiodan 4% dust @ 30 kg/hect.
   (c) Carbaryl 5% dust @ 25 kg/hect.

3. The crop may also be sprayed by any one of the following insecticide to destroy the pest:
   (a) Dimecron 85% @ 0.250 lit/hect.
   (b) Thiodan 35 EC @ 1.0 lit/hect.
   (c) Nuvan 100 @ 0.25 lit/hect.
4. 0.1% spray of brestnol 45 WP protect the crop from grubs. It acts as antifeedant due to which the larvae are unable to locate its host and finally die.

Natural Enemies

Larval Parasite: *Exacrodus populence.*

Painted Bug

Local Name: Sundar, Jhanga, Dagila Keet

Scientific Name: *Bagrada cruciferanum* Krik.

Order : Hemiptera

Sub-order : Heteroptera

Family : Pentatomidae

Host Plants: Mustard, cabbage, cauliflower, knolkhol, radish, turnip and other cruciferous plants.

Distribution: It is found throughout India and besides India distributed in Mayanmar, Sri Lanka, Iran and East Africa.

Nature of Damage: Both nymphs and adults cause the damage to the crop by sucking the sap from the leaves, stems and tender parts of the plants. The attack of pest is noted after few days of sowing the crop, if they are in great number may destroy the whole crop and necessitate resowing. The attacked plants look sickly and dry up or may get stunted in growth. The black fungus is also attracted at the feeding point due to which brown or black spots are seen on the leaves. In the later stage of the crop they suck the sap from the flowers and pods resulting in the poor formation of pods and yield. After harvesting pest may be seen in greater number on threshing floor.

Life Cycle: Three stages, *viz.*, egg, nymph and adult are found in its life cycle.

Egg: The eggs are laid by female but usually singly on the leaves, leaf stalks and stems of the host plants. Some times female also lays eggs in soil near the roots in cluster. The number of eggs laid by each female varies from 90-250. Freshly laid eggs are pale yellow in colour which turn in to pinkish at the time of hatching. The individual egg measures 1 mm x 1/2 mm. Hatching period of eggs varies from 3 to 7 days.

Nymph: The newly hatched nymph is small, bright orange in colour and with dark red eyes. They are 1.3 mm in length and 1 mm in breadth. They start feeding on the plants and develop rapidly. Nymphs moult 5 times and attain the winged adult stage. Full grown nymph measures 4.5 mm x 3.2 mm. The nymphal period varies from 20-27 days.

Adult: The adult bug is flat, bright black in colour with reddish yellow spots and dots and measures 5 mm long. The head is small, triangular and provided with black eyes, two ocelli and setaceous antennae. The scutellum is very large.

The life cycle is completed in 30-41 days and 6 to 7 generations are found in a year. It is most abundant from February to April in the plains of Uttar Pradesh. After that most of them die due to heat and only few survive in moist places.

Control

1. Since bugs congregate on the leaves and stems, they may be collected and destroyed.
2. Irrigate the field by mixing 5 kg of crude oil emulsion per hectare to destroy the hiding bugs in cracks and crevices.
3. Dusting of crop with following insecticides have been found quite effective:
   (a) Malathion 5 % dust @ 25 kg/hect.
   (b) Carbaryl 10% dust @ 25 kg/hect.

4. Spraying of any one of the following insecticides is also effective to control the bug:
   (a) Malathion 50 Ec @ 1.25 lit/hect.
   (b) Dimecron 85% @ 0.250 lit/hect.
   (c) Rogor 30 EC @ 1.00 lit/hect.

Natural Enemies: The following parasites have been found on different stages of the host.

Egg Parasite

(i) *Liophanunus samueli*

(ii) *Typhodytes sp.*

Nymph and adult parasite: Alphora sp.

Til Leaf and Pod Caterpillar

Scientific Name: Antigastra catalaunalis Dup.

Order : Lepidoptera

Family : Pyralididae

Host Plant: Til and wild varieties of *Sesamum* genus.

Distribution: The pest is found in Europe, Africa, Cyprus, Malta, Indonesia and South-East Asia. In India it is widely distributed in sesamum growing areas.

Nature of Damage: The damage is caused by the caterpillars which feed on apical shoots and young pods. The young caterpillar feeds on the leaves. They bind together the tender leaves of the growing shoot with the help of silken threads produced by them and prefer to remain within the bunch foliage thus rolled and spun together in a sort of webbed mass.

The size of the bunch become bigger and bigger as the larva grows in size and incorporates more and more of the neighbouring foliage. They also bore into the shoots, flowers buds and pods. An early attack may destroy the whole plant and single larva can destroy 2-3 young plants.

Life History: The four developmental stages *viz.*, egg, larva, pupa and adult are found in its life cycle which have been described below:

Egg: The female moths lay eggs singly generally on the underside of leaves and on the apical portions of the growing shoot: The egg are greenish or shining pale-green in colour. A single female can lay upto 150 eggs which hatch in 2-7 days depending on the season.

Larva: The newly hatched larva is 2 mm in length and some what whitish in colour; later on the colour slowly changes to green and develop black dots all over the body. The full grown larva measures 7 to 15 mm in length. The larval period lasts for 10-33 days depending upon environmental conditions.

Pupa: The full grown larvae creep to the ground and pupate in silken cocoons in the soil. Sometimes pupation takes place on the plant itself. The pupa is greenish when just formed but changes to brown later on. The pupal period lasts for 4-20 days depending upon the season.

Adult: The adult is a medium sized moth with a wing span of 15-20 cm. The fore wings are reddish-yellow in colour and there are zigzag distinct reddish decorative markings on them. The hind wings are pale-yellow or transparent. They are nocturnal in habit and remain hide during the day under fallen leaves. The moths are phototropic and are attracted to light. They live about 3 weeks and starts egg laying within 4-5 days of emergence.

The life cycle is completed in 19-64 days and there are about 14 generations in a year. All stages of the insect are seen in the autumn or even upto December. During January and February, the insect hibernates as a larva inside the stable or the harvested stalks.

Control

1. Early showing of the crop minimise its damage.
2. The infested pods and webbed leaves may be collected and burnt.

3. Spraying the crop with anyone of the following insecticides:

   (i) Chloropyrith — 0.05%

   (ii) Quinalphos — 0.05%

   (iii) Carbaryl — 0.15%

   (iv) Endosulfan — 0.07%

Hawk Moth

Scientific Name: *Acherontia styx* Westwood

Order : Lepidoptera

Family : Sphingidae

Host Plants: Til, potato, brinjal, sem and balsam etc.

Distribution: It is common in India and has been described a pest of til and is also reported from Indonesia, Sri Lanka, Philippines and Mayanmar.

Nature of Damage: The damage is caused by the caterpillar, but moth is also harmful as it sucks honey from the honey combs in apiaries. The caterpillars feed voraciously on leaves and defoliate the plants. During heavy infestation all the leaves of plants are eaten-up by the caterpillars and leaving all the stalks.

Life History: The four developmental stages namely egg, larva, pupa and adult are found in its life cycle which have been described below:

Egg: The female moths lay globular eggs singly generally on the under-surface of the leaves of the food plant and the eggs are fairly large. Their colour is at first greenish-white but later on they turn yellow. The incubation period varies from 2-5 days.

Larva: The newly hatched larva is pale-yellowish in colour. It begins to feed on the leaves and develops broad green stripes running obliquely along the dorso-lateral surface. It is a typical sphingid caterpillar with the characteristic head and caudal horn. When the larva is full grown, its colour undergoes a rather drastic change to brown.

The full grown caterpillar is about 5 cm in length and 1 cm in width, often retracts some of its anterior body segments and looks like a sphinx. The horn like projections on the hind end of the abdomen are conspicuous. The body of the caterpillar is plump and is decorated with a pleasant mixture of soft colours. The larva looks quite ferocious and venomous, although it is quite harmless and safe to handle. The larval period lasts for about 2 months.

Pupa: The full grown larva descends from the plant to the ground and finds out a suitable place for entry into the soil wherein it pupates. The pupal period lasts for 2-3 weeks.

Adult: The adult is a large reddish-brown moth with a wing span of about 10 cm. The fore wings are decorated with a mixture of dark brown and grey patterns with dark or black wavy markings and a prominent yellow spot on each wing. The hind wings have an ochre background on which there are two broad dark brown wavy cross stripes. There is a prominent Death's head mark on the thorax. They are swift fliers and often make hawk-like darts to a light source soon after dusk.

The life cycle is completed in 78 to 89 days and there are 3 generations in a year. The pest is active throughout year except winter, which is passed in pupal stage in soil. The change in colouration of the larvae and adults aid them in protective mimicry.

Control

1. Hand picking in small areas is quite effective for such large-sized insects.
2. Dusting the crop with 10% carbaryl dust @ 25-30 kg/hect.
3. Sparying the crop with 0.07% endosulfan or 0.15% carbaryl @ 750 lit/hect.

Linseed Gall-midge

Scientific Name: *Dasyneura lini* Barnes

Order : Diptera

Family : Cecidomyidae

Host Plants: Linseed.

Distribution: It is distributed in different states of India including Andhra Pradesh, Madhya Pradesh, Bihar, Uttar Pradesh, Delhi and Punjab.

Nature of Damage: The damage is caused by the maggots, which feed on the flower buds and prevent their proper opening. Consequently the seed does not set properly. The corolla of the infested flowers get crumpled and their essential organs look emaciated even of the flowers open. Due to their feeding, galls are produced and there is no pod formation. The incidence of their pest goes up to 20 per cent.

Life History: Four developmental stages, *viz.*, egg, larva, pupa and adult are found in its life cycle, which have been described below:

Egg: The female fly lays smooth, transparent eggs in the folds of flowers or in tender green buds. A female lays 29-103 eggs, either singly or in clusters of 3-5. The hatching period of eggs varies from 2 to 5 days. The length of the eggs varies from .22 to .28 mm and width range between 0.04 to 0.08 mm.

Larva: The newly hatched maggots are transparent, with a yellow patch on the abdomen. The maggots feed inside the flower-buds and eat the contents. The full grown larva becomes deep pink in colour and measures 2 mm in length. They pass through 4 instars and become full grown in 4-10 days.

Pupa: The full-grown maggots drop to the ground, prepare a cocoon and pupate in the soil. The pupal period is about 4-9 days. The female pupa is 1.9 mm long and 0.64 mm wide, while the male is smaller being 1.6 mm long and 0.5 mm wide. It is dark orange in colour.

Adult: The linseed gall-midge, is a minute (male slightly over 1 mm and female about 1.5 mm in length) and of beautiful orange colour. Thorax is dark dorsally and rest of the thorax and abdomen fuscous.

The linseed gall-midge appears in December or early in January. Its appearance is correlated with the appearance of the blossoms of linseed. The total life cycle is completed in 11-26 days and there are 4 overlapping generations in a year. The midges continue to emerge till the end of February or beginning of March depending on humidity and temperature.

Control

1. The crop should be sown easily preferable in the 1st fortnight of October.
2. The crop may be sprayed with anyone of following insecticides twice, firstly at the time of initiation on bud formation and secondly after 15 days of 1st spray:
   (i) Phosphamidon: 85 SL: 250 ml.
   (ii) Monocrotophs: 36 SL: 750 ml.
   (iii) Methyl-o-demeton: 25 EC: 1.0 lit.
   (iv) Endosulfan: 35 EC: 1.5 lit.

Natural Enemy

Systasis Dasyneura: parasite on maggots.

Cotton Leaf Roller

Common Name: Ban ki surhi, Patti mor surhi

Scientific Name: Sylepta derogata Fabr.

Order : Lepidoptera

Family : Pyralididae

Host Plants: Cotton, bhindi, sunnhemp, hollyhock and other plants of Malvaceae family.

Distribution: Besides India, it is found in Pakistan, Mayanmar, Sri Lanka, China, Japan, Jawa, Egypt, Africa and

Australia etc. In India it is distributed in all the cotton growing areas, specially Gujarat, Andhra Pradesh, Bengal, Maharashtra. Punjab and Uttar Pradesh.

Nature of Damage: The damage is done by the caterpillar. The newly hatched larva feeds on the lower surface of the leaves mainly epidermis but when it grows older rolls up the leaves in to a funnel shape and feeds from the margin. The larva remains hidden beneath the rolled leaves. Initially they are found more in number on single leaf (more than 30) but later only one is found on a leaf. Some times it appears in epidemic form and in such case 3,000 larvae have been found in 100 sq. feet (*Nag Pal,* 1948). Practically, the whole leaf is eaten up or big holes are made with the result it drops to the ground. The crop sown on the onset of monsoon and provided with nitrogenous fertilizers are subjected to much more attack of the insect. As a result of leaf damage, the photosynthesis, respiration and transpiration activities are badly affected due to which growth of the plant is affected and ultimately bears lesser bolls. The American cotton is more preferred than the Desi cotton.

Life Cycle: There are four developmental stages, *viz.*, egg, larva, pupa and adult in its life cycle which are described below.

Egg: The female moth lays eggs singly on the under side of the leaf along the thicker veins during evening hours or at night. The eggs are flat oval and white in colour which turn to brown at the time of hatching. Single female lays about 250 to 300 eggs in her life span. The hatching time varies from 2-9 days while pre-oviposition period is 2-3 days.

Larva: The young larva feeds on the lower side of the leaf and spin threads over and around it self as a protection. Later, it turns the edge of the leaf over in a fold and binds it down with silken thread, living safely within this fold. The full grown larva is green in colour with dark brown head and bears a brown coloured notch in the prothorax. At the time of pupation it becomes pinkish. It moults four times and become full grown in 15 to 35 days.

Pupa: The pupation takes place either in rolled leaves or on the ground among fallen leaves. Before changing in to pupa, the larva shrinks and become pinkish in colour. The pupa is reddish brown in colour and measures 12 mm. Pupal period lasts in about 12 days.

Adult: The adult moth is whitish in colour with faint yellow tinge, the wings with many fine dark lines forming an irregular pattern. It measures 12 mm long with wing expanse of 25 mm. Labial and maxillary palps are fused together to form a snout.

The pest over winters over as a full grown pinkish caterpillar among the fallen leaves. These larvae pupate in the second week of February, from which moths appear in March. The pest is active from March to October-November. There are 5-6 generations in a year but in Bengal only 3 generations are found. The total life cycle is completed in 25 to 54 days depending upon environmental conditions.

Control

1. The field should be ploughed deeply before sowing so that hibernating larvae may be destroyed.
2. Regular hand picking of rolled leaves minimise the attack of pest.
3. Balance nitrogen should be given and if possible resistant varieties, *viz.*, 10 F should be preferred.
4. The crop may be dusted with any one of the following insecticides to control the pest:
   (a) Carbaryl 10% dust @ 20 kg/hect.
   (b) Fenitrothion 5% dust @ 20 kg/hect.
5. Effective control can also be achieved by spraying the crop with any one of the following insecticides:
   (a) Carbaryl 50 WP-3.00 kg/hect.
   (b) Endosulfan 35 EC-2.50 lit/hect.
   (c) Formothion 25 EC@ 1.00 lit/hect.

6. Spray of 0.1 percent brestenol 45 WP protects the crop. It acts as antifeedant.

Natural Enemies: The following are the effective natural enemies of the pest:

Egg Parasite: *Trichogramma minutum, R.*

Larval Parasite

(a) *Bessa remota,* Alder.

(b) *Microbracon lefroyi* D.

(c) *Brachymeria tachardiae,* Cam.

(d) *Xanthopimpla punetata,* F.

(e) *Microtoridea lissonara,* Vier.

(f) *Elasmus indicus,* Rohw.

Pupal Parasite: *Xanthopimpla sp.*

Many birds also prey on the larva of the pest which may be encouraged.

Spotted Boll Worm

Common Name: Chitkabri surhi

Scientific Name

(i) *Earias vitella (fabia),* Stoll.

(ii) *Earias insulana,* Boisd

Order : Lepidoptera

Family : Cymbidae

Host Plants: Cotton, Bhindi, Sunnhemp, Hollyhock and other Malvaceous plants.

Distribution: It is found in Spain, Syria, Palestine, Egypt, India, Mayanmar, Madagascar and Australia. In India it is commonly occur in M.P., Maharashtra, Mysore, Chennai, Gujarat, U.P., Delhi and Punjab etc.

Nature of Damage: The larva is the destructive stage of the pest which in the beginning (when plants are small 20-30 cm high) bores the top portion of the growing shoot which withers and drops.

As a result of which plant growth is adversely affected. *Lefroy* (1906) stated that this pest is more serious than locusts. When the flower buds and bolls appear the larvae start feeding on them. The caterpillars bore in to bolls close the hole with their excreta and feed inside.

The attacked bolls are generally shed those left on the plants open prematurely with the result poor quality of fibre is produced and market value is reduced, Several holes plugged with excreta can be seen on the bolls and a single larva may destroy many bolls in its life time. The damage caused by the pest is estimated to the tune of 20-40% by Claiston (1921) and 50% by Mishra (1938).

Life History: There are four developmental stages namely egg, larva, pupa and adult in its life history which are described as follows:

Egg: The female moth lays eggs generally singly, scattered over fresh squares etc. The eggs are spherical, bluish green in colour with parallel longitudinal ridges which projects upward and give it crowned appearance. The female lays 200-500 eggs in her life time only at night. An individual egg is 1 mm in diameter. The hatching period of eggs varies from 3-4 days in summer and 8-9 days in winter.

Larva: The newly hatched larva is brownish, white with a dark head and prothoracic shield. It measures 1.2 to 1.3 mm in length. The caterpillar remains about 24 hours outside the shoots or bolls and after that it enters in to the shoot or boll. Larva moults 3 or 4 times become full grown measuring 22-25 mm long.

The caterpillar of *E. vitella* is brownish with a median longitudinal streak while of *E. insulana* has dull greenish body with a number of black marks and oval, orange dots on the prothorax. The caterpillar becomes full grown in 10-12 days during summer and 15-20 days during winter.

Pupa: The full grown caterpillar comes out of the bolls and pupates either on the plant or on the ground among fallen leaves or 5-25 cm deep in soil. The pupa is enclosed in tough silken cocoon of dirty white or light brown colour. It measures 12 mm in length and 4-5 mm in breadth. Pupal period lasts for 5-15 days during summer and rainy season and 18-31 days during winter.

Adult

Earias Vitella: The moth measures 12 mm long and 21 across the wings, the fore wings have a broad green band extending from the base to apex and hind wings are white in colour. The body is pinkish white in colour having tuft of hairs behind the thorax.

Earias Insulana: It measures about the same as above, but fore wings are completely green. The species is more prevalent in dry regions.

The pest remains active throughout the year, but population is low during December to March. It begins to multiply in April-May and reaches its maximum during July to September. The life cycle is completed in 31-55 days and 7-8 generations are found in a year. The males die after copulation while females survive 10 to 20 days depending upon the season.

Control

1. Destruction of plants which harbour the pest.
2. After harvesting of crops the field should be ploughed to destroy the hidden caterpillars.
3. Collection and destruction of infested shoots and bolls etc.
4. The crop may be sprayed with any one of the following insecticides to minimise the attack of pest:
   (a) Thiometon 25 EC @ 1.25 lit/hect.
   (b) Endosulfan 35 EC @ 2.5 lit/hect.
   (c) Fenetrothion 50 EC @ 3.00 lit/hect.

(d) Phosphomidan 85% @ 500 EC lit/hect.

(e) Carbaryl 50 WP @ 2.5 kg/hect.

The crop should be sprayed twice first at the time of hatching of the eggs (peak hatching period) and second after 15 days of previous spraying. The quantity of water may be required 700-750 lits/hect. If the insecticides are to be applied on bhindi crop endosulfan or D D V P should be preferred.

Natural Enemies: The following are the natural enemies on different stages of the pest:

Egg Parasite: *Trichogramma spp.*

Larval Parasite

(a) *Microbracon lefroyi, D* and G

(b) *M. Hebets, Say*

(c) *Elasmus sp.*

Pupal Parasite

(a) *Melcha mursei,* Carn.

(b) *Chelonus rufus,* Lyle.

(c) *Chalchis* sp.

Pink Boll Worm

Local Name: Kapas ki Lal Surhi

Scientific Name: *Pectinopliora (Platyedra)* gossypiella Saund.

Order : Lepidoptera

Family : Gelechiidae

Host Plants: Mainly cotton, but some time on Deccan Hemp, Kanghi and Gulkhara etc.

Distribution: It is widely distributed and probably the most serious cotton pest on a world basis. Besides India, it is mainly found in Mayanmar, Sri Lanka, Egypt, South Africa, Maxico, Brazil, Hawaii, Philippines and Australia. In India it was first of all found in Mumbai (Maharashtra) during 1842 and now distributed in U.P., Punjab, Rajasthan, Andhra Pradesh, Gujarat, Bihar and other cotton growing states.

Nature of Damage: The damage is caused by the larvae which bore in to flower buds, flowers and young bolls. The young attacked bolls are invariably fall down. The pink boll worm larvae do most spectacular damage to practically mature cotton bolls in which they enter at such a tiny stage (just-hatched larvae) that their entry hole gets healed and in which they remain devouring both seed and fibre-forming tissues.

The infestation some times is so severe that up to 10 caterpillars may be found in a single boll and 75 to 100 per cent bolls may be found infested. The oil content and the germinating capacity of the seed and lint formation are badly affected. The American cotton is more damaged than the Desi cotton and may cause 25-50% loss.

The larva of spotted boll worm prefer to eat the gynaecium while pink boll worm eats the anthers of the flowers, the holes on big bolls are plugged by the former while remain open in case of later.

Life History: There are four developmental stages, *viz.*, egg, larva, pupa and adult are found in its life cycle which are described as below:

Egg: The female moth starts egg laying after 1 or 2 days of copulation. The eggs are laid singly or in batches of 2-10 on leaves, buds, flowers and bolls. A single female can lay 125 to 300 eggs which are 0.50 mm in length and 0.30 mm in breadth. The eggs are flattened, elongated and nearly white when laid freshly, becoming brown on the third day and deep brown at the time of hatching. The hatching period of eggs varies from 4 to 29 days depending upon the environmental conditions in U.P. during rainy season, it hatches in 4-5 days.

Larva: The newly hatched larva is colourless and about 1 mm in length. It bores in to the boll, begins to feed one and when one seed is finished, it passes on to the next. The entry holes get quickly healed up and it is difficult to distinguish an infested boll from an uninfested one. The full grown caterpillar is 13 x 2 mm long and pinkish in colour. The larvae undergo generally 3 moults and the larval period depends on two distinct types of life cycles referred as short life cycle and long life cycle. The short life cycle is completed just like other lepidopterous pest in 30 to 90 days.

In case of long life cycle, the larvae enter in to resting stage where they remain over a period of 8-12 months and some times up to 2 years before they pupate and moths emerge. The interesting point in the long life cycle is that at certain times of the year two types of larvae are found in a mixed population. One type pupates in the normal manner, while the other type makes preparations for prolonged hibernation.

The larva scoopes and binds together around it self generally two but some time more than two seeds with the help of silken threads and then enters the resting stage in the hibernaculum prepared between double seeds. It is this stage which has been responsible for the universal distribution of the pink boll worm in the cotton growing regions of the world. When this double-seed is sown along with healthy seeds, the resting larva comes out of its hibernaculum and makes a tunnel up to the surface and then goes down to pupate.

Pupa: The pupation takes place in a silken cocoon on the ground among fallen leaves and flowers etc. The pupa is light brown in colour and measures 6 mm in length and 2.6 mm in breadth. At the time of hatching the colour of pupa changes in to dark brown. The pupal period lasts in 6-20 days.

Adult: The adults are small moths of dark, fuscous brown colour and about 1 cm in body length. The antennae are filiform, the palpi long and curved. The fore wings are pointing towards margins and possesses long brown fringes. They are active during the night.

The moths are comparatively quite long-lived. Male generally survives up to 20 days, while female up to 55 days.

Under normal conditions, the life cycle is completed in 22 to 77 days but the long-life cycle is completed in about 13.5 months in U.P. There are 4 to 6 generations in a year.

Control

1. Collection and destruction of infested flowers, flower buds and bolls etc.
2. Heat all cotton seeds at about 104-110F., either in the sun or by simon cotton seed heater to kill the hibernating larvae in them.
3. Seeds should be fumigated, either with methyl bromide @ 3.5 kg/100 sq. m seeds for 24 hrs or with aluminium phosphide @ 45 tablets/100 sq. m seeds for 7 days before sowing.
4. Ratooning of cotton should not be practiced in infested areas and all other alternate hosts should be destroyed.
5. The spraying may be done with any one of the following insecticides at fortnightly intervals twice or thrice to destroy the pest:
   (a) Thiometon 25 EC @ 1.000 lit/hect.
   (b) Phosphomidon 85% @ 0.500 lit/hect.
   (c) Endosulfan 35 EC @ 2.500 lit/hect.
   (d) Fenitrothion 50 EC @ 3.000 lit/hect.
   (e) Carbaryl 50 WP @ 2.500 kg/hect.

Natural Enemies: The following are natural enemies found in different stages of the pest.

Egg Parasite

(a) *Triphlapi pectinophorae*

Larval Parasite

(a) *Microbracon lefroyi* (D and G)
(b) *M. greeni*
(c) *Bracon* kitchneri (Will)

(d) *Apanteles pectinophorae*

(e) *Elasmus pus ptyedrae* Ferr.

Legal Control: As per direction of Director of Plant Protection, Quarantine and Storage, Ministry of the Food and Agriculture India, all imported cotton must be fumigated at ports.

Gram Cut Worm

Common Name: Kata, Kirauna, Kamta, Katuwa, Surhi

Scientific Name: *Agrotis ypsilon* Rott.

Order : Lepidoptera

Family : Noctuidae

The other two spp. are as follows:

Agrvtis flammatra Schiff.

A. segetum Schiff.

Host Plants: Gram, Potato, Pea, Wheat, Tobacco, Cotton, Mustard, Groundnut, Bhindi, Lucerne, Cabbage etc. Thus they are polyphagous in nature and attack many crops.

Distribution: Besides India, it is distributed in America, Africa, Sri Lanka, Mayanmar, New Zealand, Malaya and European countries. In India the cut worms are found in all the states with particular reference to U.P., Bihar, Rajasthan and Madhya Pradesh.

Nature of Damage: In India winter is generally lean season for insect-pest activity, but the cutworms belong to that small group of insects, the destructive activity of which is more marked in the rabi crops. They are known as cutworms because these cut the plants and fell down to the ground either the whole plants like gram or their twigs. They are some times also called as surface caterpillars because their activities are mostly confined few centimetres of the surface soil. The damage is caused by the larvae which hide during the day and come out in the evening to cause the damage. First of all they feed on the epidermis of the fallen leaves or the green leaves touching the ground. Later, the caterpillars cut the leaf or shoot or the plants just above the ground level and burried in the soil. They destroy many more plants than they actually feed upon and thus cause a very serious loss. The attack of cut worm in gram is confined from November to February.

Life History: The four developmental stages, viz., egg, larva, pupa and adult are found in its cycle which are described as below:

Egg: The moths of the cut worms are strong fliers. They come to the plains in winter from hills and return in summer to hills again. After mating, the females lay eggs mostly on the underside of the leaves or on the soil near the root base. The eggs are laid singly or in clusters of up to 30 eggs. A female generally lay 300-350 eggs but can lay thousands of eggs. The eggs are round yellowish in colour and measuring 1 mm long. The hatching period of eggs varies from 2-6 days but during severe cold 12-15 days.

Larva: The young larvae feed on their egg-shell as their first meal and fall to the ground at the slightest disturbance. They are light bluish-green in colour and measuring 2 mm long. The caterpillars remain hidden in the loose soil during the day and come out at night. The full grown caterpillar is about 40-45 mm in length, plump, smooth and flattened in look and dull grey brown in colour. The caterpillar becomes full grown in 21-36 days.

Pupa: The pupation takes place in the soil. The full grown caterpillar enters the soil much deeper than during its usual activity and constructs the earthen chamber, the sides of which are made smooth and hard by an exudation from the larval mouth. The pupae are reddish brown in colour and about 20 mm in length and 7 mm in breadth. The pupal stage lasts for 10-30 days.

Adult

Agrotis Ypsilon: The moths are brown in colour measuring 25 mm in length and 7 mm in breadth with a wing expanse of 40-50 mm. The males have bi-pectinate antennae while females have filiform. Fore wings are dirty white in colour having dirty brown spots while hind wings are white. The terminal end of abdomen bears a tuft of hairs.

Agrotis Flammatra: The fore wings are characteristically marked and two third of costal area is pale, grey brown having a kidney shaped and a semicircular spot below the pale area and two black wavy lines near the base. A black sign of inverted V (^) is found on prothorax.

Agrotis Segetum: It is mostly found in hills and smaller than the above two species. The colour of the body is brownish-orange.

They are active from October to March in plains and after which most of them migrate to hills. The life cycle is completed in 35 to 84 days and generally two generations are found in plains and in the hills four generations occur during summer.

Control

1. The moth may be destroyed after attracting them in light-traps.
2. Hand-picking of larvae may also reduce the population of caterpillars.
3. Flooding of field may kill the caterpillars.
4. Any one of the following dusts may be mixed in the soil at the time of sowing to kill the larvae of the pest:

   (a) Methyl parathion 2% dust @ 25 kg/hect.

   (b) Heptachlor 3% dust @ 35 kg/hect.

   BHC and aldrin dusts should not be used in tuber crops.
5. Any one of the following insecticides may be sprayed near the base of plants in standing crop to control the pest:

(a) Quinalphos 25 EC @ 3.5 lit/hect.

(b) Gamma BHC 20 EC @ 3.75 lit/hect.

(c) Chlorpyriphos 20 EC @ 3.5 lit/hect.

Natural Enemies: The following insects are the larval parasite of the pest:

(a) *Braconis* wasp

(b) *Apanteles* sp.

(c) *Microbracon* sp.

Gram Caterpillar

Common Name: Chane ke illi or surhi, Fali bedhak.

Scientific Name: *Helıcoverpa (Heliothis) armigera* Hubn.

Order : Lepidoptera

Family : Noctuidae

Host Plants: Gram, Pea, Cotton, Arhar, Maize, Cobs, Tomato and Tobacco etc.

Distribution: It is widely distributed and practically omnivorous pest. In United States of America, it is a serious pest of cotton and is known as the cotton boll worm. In India it is mostly found in Punjab, U.P., Rajasthan, Bihar and Madhya Pradesh.

Nature of Damage: The pest is active through out the year but damage to gram is cause from November to March. The damage is caused by the larvae which feed on the leaves and destroy the seedlings in the early stages.

At the time of pod formation it is found feeding on the developing grain after cutting a hole in the pod and thrusting its heads therein. It does not hide in the soil during day but remains hide in the plants. It has been estimated that a single caterpillar destroys 30-40 pods of gram in its life time. It has been observed that grown caterpillars feed on the small larvae of its own which is known as cannibelism. In severe infestation damage may be caused from 20-50 per cent.

After the harvest of gram crop in March-April it migrates to tomato and feeds on ripening fruits which invites the roting of the fruits.

Life History: Four developmental stages, *viz.*, egg, larva, pupa and adult are found in its life cycle which are described as below:

Egg: The adults moths are visible in the gram field during early part of November which are active after dusk. Two to four days of copulation female lays light yellow eggs generally on the side of the leaves or on twigs or on pods. The eggs are spherical, sculptured and with greenish shine. A single female can lay 500 to 1000 eggs in her life span. The egg-period lasts from 3 to 7 days.

Larva: The newly hatched larva measures 2 to 2.5 mm long and light-green in colour. The young larva begins to feed on the tender portions of the leaves and shoots and when pod formation takes place it feeds on the pods. The full grown caterpillar is about 35 mm in length and very variable in colour. Generally it is pale green with a dark mid-dorsal and two yellow lateral stripes. The caterpillar moults four times and becomes full grown in 20-25 days.

Pupa: The larva when full grown drops to the ground and pupate in a shell 8 to 10 cm deep in the soil. The pupa is dark brown in colour and about 16 mm in length. It remains in the soil during the cold months of January and February. The pupal period generally lasts from 10 to 24 days depending upon the environmental conditions.

Adult: It is a typical stout-built noctuid moth with a wing span of 30-40 mm, and colour varying from olive grey to red brown. The fore wing has dark specks and dark area near the outer margin. The hind wings are white with dark veins and with a broad blackish band along the outer margin. There are filiform antennae in males and bipectinate in females. The moths of *H. armigera* are 20-25 mm in length while that of *H. asulta,* 16 mm in length with a wing span of 37 mm.

The moths emerge in late February or early March from over wintering pupae and lay eggs again. The total life cycle is completed in 35-70 days on gram.

Control

1. Deep ploughing of infested field in the summer to expose the pupae is found useful.
2. The adult moths may be attracted in light-traps and destroyed. Any one of following dusts may be used to destroy the pest:
   (a) Malathion 5% dust @ 25 kg/hect.
   (b) Carbaryl 10% dust @ 25 kg/hect.
   (c) Fenitrothion 5% dust @ 20 kg/hect.
   (d) Phenthoate 2% dust @ 25 kg/hect.
4. Spraying with any one of the following insecticides has also been found effective to control the pest:
   (a) Endosulfan 35 EC @ 1.25 lit/hect.
   (b) Carbaryl 50 WP @ 2.50 kg/hect.
   (c) Gamma BHC 20 EC @ 2.5 kg/hect.
   (d) Monocrotophos 36 SC @ 1.00 lit/hect.

The above quantity of insecticide may be mixed in 700-750 tit of water and sprayed on the crop.

Natural Enemies: The following are the important natural enemies:

Egg Parasite

(a) *Trichogramma minutum*

(b) *Triphleps insidious*

Larval Parasite

(a) *Apanteles sp.*

(b) *Bracon sp.*

(c) *Chelonus narayani*

Pea Leaf Miner

Scientific Name: *Phytomyza atricomis* Meigen

Order : Diptera

Family : Agromyzidae

Host Plants: It is polyphagous pest mostly feeds on pea, crucifers, Unseed and potato.

Distribution: Besides India, it is found in Europe, America, Russia, Australia and New Zealand etc. In India, it has been recorded from 29 different species of economic plants belonging to 13 families from different states.

Nature of Damage: The larvae are the damaging stage which make prominent whitish tunnels in the leaves. If the attacked leaves are held against bright light, minute slender larvae an be seen feeding within the tunnels. The huge number of tunnels made by the larva interfere the photosynthesis and the proper growth of plants. The female fly also punctures the tender leaves with its ovipositor and then turns round to feed on the plant juice exuding from the puncture.

Each fly makes numerous such punctures of each leaf and these injured points develop into as many protuberances of the leaf *Saxena* and *others* (1969) recorded a severe infestation of the leaf minor ranging upto 85.7 per cent.

Life History: The four developmental stages, *viz.*, egg, larva, pupa and adult are found in its life cycle which are described below:

Egg: The flies emerge in the beginning of December and after mating start egg laying singly in the leaf tissues. The eggs are very small, about 0.3 mm in length, and oval elongate in shape. A female fly lays 350-400 eggs and the incubation period is 3 to 7 days.

Larva: The tiny maggots after hatching from the eggs begin to feed on the mesophyl of the leaf without damaging the two epidermal layers. Maggots remain within the tunneled gallaries and become full grown in 4-10 days. The full grown larva is about 3 mm long and light pinkish in colour.

Pupa: The pupation takes place within the gallaries where larva pupates within its last larval shin which hardens in due course. The pupa becomes reddish-brown or dark-brown in colour. The pupal period ranges from 6 to 12 days.

Adult: The adult is very small fly, about 1.5 mm in length and with a wing span of 3.5 mm. It is two winged, which has a greyish-black mesonotum and yellowish frons.

The pest is active from December to May and from May onwards it passes the rest of the year in soil in the pupal stage. The total life cycle is completed in 14-21 days and 4-5 generations are generally found in a year.

Control

1. Collect the damaged portions of the plant during initial stage and destroy to minimise the multiplication.

   The pest can be controlled by repeated spraying (after every 2-3 weeks) of anyone of the following insecticides:

   (i) Dimethoate : 0.03%
   (ii) Formothion : 0.03%
   (iii) Carbaryl : 0.1%
   (iv) Monocrotophos : 0.03%

Pea Bod Borer

Scientific Name: *Etiella zinckenella*, Treitscke

Order : Lepidoptera

Family : Phycitidae

Host Plants: Pea, lentil, red gram, lathyrus and sunnhemp.

Distribution: It is distributed throughout India with particular reference to Uttar Pradesh, Bihar, Madhya Pradesh and Punjab.

Nature of Damage: Newly hatched larvae constructs a web type structure through which they entered the pod. After 2-3 days of feeding inside the seed, the caterpillar come-out to feed superficially on the seeds. They migrate from one seed to another inside the pod without completely consuming the damaged seeds. In the later stage of damage, the pods become discoloured, watery in appearance and emitted foul smell. On an average 77.6 per cent of seeds in infested pods are found damaged by the pest. The yield of pea-seeds is reduced on an average by 23.9 per cent due to the feeding of the larvae.

Life History: The four developmental stages namely egg, larva, pupa and adult are found in its life cycle which are described below:

Egg: The eggs are laid on blossoms, buds, twigs and leaves of the plants. They are generally deposited singly but sometimes also in batches of 4-6 eggs. A single female lays 44 to 178 eggs in her life span. The incubation period varies from 4 to 7 days which may extend up 33 days in adverse conditions.

Larva: The newly hatched caterpillar is yellowish in colour which changes to milky white after 10-12 hours of feeding. The colour further changes to light green after 6-7 days of feeding and ultimately it becomes pinkish when full-fed. The larva bores into the pod and feeds on the grain inside. Often more than one larva is observed in a pod. The larval period lasts for 10-17 days. Cannibalism has also been observed amongst young and between young and older caterpillars..

Pupa: Full grown larva comesout of the pod and pupates in soil. Pupation takes place in cocoons in the soil at a depth of 2 to 4 cm. Pre-pupal stage ranges from 2-4 days. Pupal period lasts for 10-20 days depending upon the temperature.

Adult: The emergence of moth takes place at night and in a few cases in the early hours of the morning. The moths are grey in colour with wing expanse of 25 mm. The fore wings have dark marginal lines and are interspersed with ochreous scales. The moths emerge in February and March nocturnal in habit.

The pest breeds throughout the year. The total life cycle is completed in 27 to 46 days and passes through 5 generations. The moths start appearing in the fields in during last week of February. All the stages of the pest are available in the field during the months of March and April.

Control

1. Spraying the crop with anyone of the following insecticides per hectare:
   (i) Methyl demeton 25 EC: 1.50 litres.
   (ii) Endosulfan 35 EC: 1.5 litres.
   (iii) Carbaryl 50 WP: 2.25 kg.
2. Dusting of any one of the following insecticides has also found effective:
   (i) Endosulfan 2% dust: 25 kg.
   (ii) Carbaryl 5% dust: 25 kg.
   (iii) Fenitrothion 5% dust: 25 kg.

3

Destroyers of Fruits

Different Insects

Mango Mealy Bug

Local Name: Safed phunga, Gujia, Aam ka matkun.

Scientific Name: *Drosicha mangiferae (stebbingi)* Green.

Order : Hemiptera

Sub-order : Homoptera

Family : Coccidae

Host Plants: The list of food plants includes about 62 species of trees, shrubs and herbs with particular reference to mango, guava, peaches, plump, rose, castor, jack fruit, papaya, fig and lemon etc.

Distribution: The pest is distributed in India, Pakistan, Sri Lanka, Mayanmar, Malaya, China and Formosa etc. In India, it is mainly found in Assam, Bihar, Delhi, Chennai, Orissa, Punjab, Uttar Pradesh and Madhya Pradesh.

Nature of Damage: The damage is caused by the nymphs of both sexes and female adults which suck the sap from twigs, shoots and flowers. As a result of which flowers dry up and only few fruits are set. The fruit bearing twigs are so weak that with the slightest jerk whether by bird or wind, the fruits fall down. They secrete honey-dew which attracts fungus due to which black spots may be seen on the twigs and shoots etc. The attack is much more serious during spring when shoots are richly supplied with overflowering cell sap. The attack persists from November to April-May and by that time it changes in to adult. The damage caused by this pest is estimated to the tune of 20-50 per cent depending upon the severity of attack.

Life Circle: The female and male adults develop differently. In the case of female 3 developmental stages, *viz.*, egg, nymph and adult are found, while in case of male egg, nymph, pupa and adult are found.

Egg: The adult bugs are common from the end of March to the middle of May. The adult gravid females after fertilization crawl down along the tree trunk to the ground or some times just fall down. It lays eggs in the soil at a depth of about 8-15 cm in clusters of about 300-400 eggs which are enclosed in a sac-like pouch of some silken material, after 15-16 days of copulation. The eggs in such a cluster are deposited by the female in instalments over a period of 7-15 days. The oviposition is generally confined to an area of a few feet in diameter round the base of the tree. The eggs are cylindrical pinkish-brown in colour and measuring 1.25 x 1.00 mm long. The eggs laid generally during April and May remain in the soil till the end of November or early December, when they start hatching.

Nymph: The newly hatched nymphs are red in colour and may be seen crawling up the tree trunk. The 1st instar nymphs of both the sexes can not be differentiated. The 1st instar nymphs are 1 mm long and moult in the month of January about two months after hatching. The IInd instar nymphs on the basis of length may be differentiated; female nymphs being larger of the male. The development of both the, sexes takes place as follows:

Development of Female Nymph: The second instar nymph moults in the month of March about 1 and 1/2 months of 1st moult and measures 8 mm long at this stage. Initially they are brown in colour and after some times they are covered with white powder resulting in to white in colour. They grow gradually and become 1 cm long. In the mean time reproductive organs are developed. Some times they moult three times, then reproductive organs developed in fourth instar.

Development of Male Nymph: In the case of males, after first moulting a short prepupal and pupal stages are noticed. The pupation takes place in the soil on the fallen leaves in cocoons made by waxy thread secreted by nymphs. Pupal period is of about 21 days.

Adult: There is a well established sexual dimorphism in this stage which is found from April to June.

Male: The male is the winged insect with only one pair of wings and a very delicate reddish body. It measures 5 mm x 2 mm in length and breadth, respectively with a wing span of 13 mm. The antennae are whorled type and black in colour. They live about a week and having no mouth parts hence do not feed.

Female: The female is wingless fleshy flat bodies insect with a length of 14-16 mm and a breadth of 8.5 mm. Its body is covered with ashed white mealy powder having 3 pairs of small black legs. It has piercing and sucking type of mouth parts. The female moth survives up to a month.

The life cycle is completed in about 9.5 to 11 months and only one generation is found in a year.

Control

1. Raking of the soil around the base of the infested tree so that egg masses get exposed to the sun and may be killed by heat in the month of May and June.
2. Application of sticky band around the tree trunk so as to check the nymphs from crawling up the trees

(4 parts castor oil + 5 parts resin) 1/2 to 1 meter above the ground level during the second week of December. It remains effective for a period of two weeks after which it should be repeated.

3. 2% methyl parathion dust may be dusted in the soil around the tree trunk in 1 meter diameter and also on the trunk up to height of 1 meter.
4. To kill the nymphs which are on the tree any of the following insecticides may be sprayed in a tree after mixing in 15-20 litres of water.
   (a) Phosphamidon 85% — 6 ml.
   (b) Malathion 50 Ec — 50 ml.
   (c) Endosulfan 35 Ec — 20 ml.
   (d) Formothion 25 Ec — 30 ml.

Natural Enemies: Some Chalcid wasps and *Auliis vestila* have been found parasitising the nymphs.

Mango Leaf Hopper

Local Name: Aam ka Chenpa, Pharka, Lassi Thala, Tela

Scientific Name: *Amritodus (Idiocerus) atkinsoni* Leth

Other Species: *Idiocerus clypealis* Leth.

I. niveosparsus Leth.

Order : Hemiptera

Sub-order : Homoptera

Family : Jassidae

Host Plant: Mango.

Distribution: It is found in India, Sri Lanka, Mayanmar, Malaya, Formosa and Australia etc. In India it is commonly found in Bihar, Uttar Pradesh, Punjab, Mumbai, Andhra Pradesh and Madhya Pradesh.

Nature of Damage: The damage is caused both by the adult hoppers and their nymphs in way of sucking the sap from the new shoots, buds and flowers, resulting the entire inflorescence and even small fruits to dry up, nymphs cause more damage than the adults as they are devoid of wings hence can not fly, therefore, suck the sap from a particular place for long time. They also secrete honey-dew which interferes with the respiration of the leaves. The heavy infestation of the pest usually occur periodically after every three or four years that is why the pest is known as periodic insect pest. It is suspected that one of the causes of the irregular bearings in mango is the periodical attack of this pest. Its attack starts from the end of the March and continue till the end of June. The old trees are more damaged than the newly planted trees.

Life History: Three developmental stages namely egg, nymph and adult are found in its life cycle which are described as below:

Egg: The adult hoppers hibernate in winter and become active in the beginning of February. After 3 days of copulation female lays eggs singly in the tissues of leaves, buds and flowers. The eggs are cylendrical dull white in colour and measure 0.9 mm, long. A female can lay 200 eggs in her life-span. The eggs hatch in 7-10 days.

Nymph: The newly hatched nymphs are yellowish green in colour and measure 1 mm long. They suck the sap from the tender shoots, flower buds and flowers and during this period they move about on the tree but do not fly. They moult 5 times grow in size and after 3-4 weeks of feeding they become adult.

Adult: The insect is triangular in shape being broader at anterior end and narrower towards posterior. It is 5 mm long and dull brown in colour. They cluster on the stems and thicker branches. *Idiocerus clypealis* is smaller than the above and generally rests on the under surface of the leaves. The adults of both the species have two pairs of transparent wings and 3 pairs of legs. The anterior ones are smaller and poserior ones are larger.

The life cycle is completed in 31-41 days. There are two generations in Northern India. The adult hoppers crowd on the stems and leaves, mostly on the under surface and fly about in all direction on the disturbance from middle of April to June. Then-population decline in rainy and winter seasons. They prefer shady and damp places for hide. They become again active in the commencement of spring.

Control

1. The Plants should be planted at proper distance and garden should be open to maximum sunlight and air by cutting unwanted branches trees and bushes.
2. The plants may be sprayed with any one of the following insecticides. The given amount of insecticide may be mixed in 15-20 litres of water and sprayed on a tree.
   (a) Phosphamidon 85%-6 ml.
   (b) DDVP 76 EC-10 ml.
   (c) Malathion 50 EC-50 ml.
   (d) Endosulfan 35 EC-20 ml.
   (e) Formothion 25 EC-30 ml.
   (f) Carbaryl 50 WP-25 g.
   (g) Dimethoate 30 EC-25 ml.

Natural Enemies: The following are the natural enemies of the pest:

Nymphal Parasite

(a) *Pipunculus sp.*

(b) *Epipyrops sp.*

Adult Parasite: *Dryinus sp.*

Bark Eating Caterpillar

Scientific Name: *Inderbela quadrinotata* Walk. *I. tetraonis* Moore

Order : Lepidoptera

Family : Tetragridae

Host Plants: Mango, guava, citrus, pomegranate, jamun, ber, litchi and rose etc.

Distribution: The pest is widely distributed in India, Mayanmar and Malaysia. In India it is found all over the country with particular reference to the Vidarbha region of Maharashtra.

Nature of Damage: The caterpillar eats the bark and bores in the stem near the junctions of branches. It is very harmful to the tree as it feeds on the bark and in this process it seriously injures the plant vessels through which nutritive plant-sap is transported within the plant's system; the result is that the tree-growth and fruit bearing capacity are adversely affected. Sometimes the infested branches can dry up and in cases of severe infestation, the whole tree may die. The dirty elongated zig-zag ribbon-like messy web consisting of bits of bark pieces, excreta, etc. covering the trunks and branches of trees may be seen. They prefer older trees than younger ones. The infestation in some areas of the country has been reported to as high as 40 per cent.

Life History: The four developmental stages namely egg, larva, pupa and adult are found which are described as below:

Egg: The egg laying takes place in May and June. The female moth lays a very large number of eggs in small groups of 15-20 eggs in each. They are laid in crevices of bark and incubation period is about 10 days.

Larva: The larva soon after hatching begins to feed on the bark and prepares the web under which it lives. They remain hidden during the day but come out at night for feeding on the bark and they do so under coverings of webbings containing frass, faecal pellets etc.

The larval period is quite long and the caterpillar continues its destructive activity from May-June to April the following year. During this period, it grows slowly being about 1.5 cm in September and about 4 cm by December when it is practically full grown. Thus, the larval period extends over several months.

Pupa: The pupation starts from April onwards. It takes place within the larval gallery. At the end of pupal period, the pupa wriggles up to the opening of the gallery where from the moth emerges, leaving the pupal skin protruding from the exit hole. Pupal period lasts 3-4 weeks.

Adult: The adult is large-sized moth with a wing-span of about 4 cm in the female and about 3 cm in the male. It is light grey to light brick-red in colour and with dark brown patches or dots. The moth's life is quite short and after finishing egg-laying within 2-3 days, it dies.

The life cycle is completed in a year, therefore, only one generation is found per year.

Control

1. Clean the web and fumigate the hole with an ordinary fumigant by putting in the hole a swab of cotton wool dipped in the liquid fumigant (petrol, kerosine, carbon-di-sulphide) and inserting the same in the whole which should thereafter be plugged with mud.
2. Drenching the tree trunks and branches with any one of the following insecticides in the month of June:

(i)	Endosulfan	: 0.05%
(ii)	Carbaryl	: 0.15%
(iii)	Telodrin	: 0.03%
(iv)	Chlorpyrifos	: 0.03%

Mango Stem Borer

Scientific Name: *Batocera rufomaculata* Dejean.

Order : Coleoptera

Family : Cerambycidae

Host Plants: Mango, jackfruit, fig and mulburrey etc.

Distribution: The pest is widely distributed throughout India.

Nature of Damage: Damage is caused by the grubs. They tunnel into the branches and trunks of the tree and feed on the wood inside. Sometimes they also attack the plant near ground level and even on the roots. The tunnels, which are 3 cm or more in cross-section, are packed with coarse chips of wood and fibre. Though the external symptoms of attack are not always visible, the site can be located from the sap on frass that comes out of the hole. During severe infestations, the branches are made completely hollow and therefore, easily break off or dry up.

Life History: The four developmental stages, *viz.*, egg, larva, pupa and adult are found in its life cycle which are described below:

Egg: The beetles generally appear during the monsoon and deposit eggs in the bark of trees, specially such trees which are dead or unhealthy or have wounds or cuts in them. The hatching period varies from 5-8 days.

Larva: The newly hatched grubs bore into the wood and make irregular channels inside. The grubs possess strong biting mouth parts and penetrate into the stem or even the roots and feed on woody tissues. The winter is passed in the grub stage inside the burrow. When winter is over, they again begin to feed in the spring. The full grown larva is leg less, stout, yellowish-white, fleshy and measuring about 6 cm in length. Its head is dark, with well developed mandibles. The larval stage lasts for 10-12 months and sometimes even more than this.

Pupa: The full grown grubs hollow-out a cell for pupation inside the stem. The pupal period lasts for 3-4 weeks and after which the adult beetles emerge from circular exit hole in the trees.

Adult: The adult beetle is about 5 cm in length and 2 cm in breadth. It has long horn and pale-greyish in colour, with several creamy, irregular spots on its back. The head is distinct, with large prominent eyes, and the pronotum is ornamented with two crescent orange-yellow spots. The beetles are nocturnal in habit and feed to some extent on the bark of the living twings etc.

The life cycle is completed in 12 months and only one generation is found in a year. During the non-active season, the pest takes shelter under the bark of the mango tree or cracks and crevices in the adult stage. Adult beetles generally appear in May and June.

Control

1. Prune and destroy all the affected branches during winter.
2. In case of stems, insert the soacked cotton in fumigants like carbon-di-sulphide etc. in the holes.
3. Drench the tree trunks and branches with 0.05% endosulfan or 0.03% telodrin emulsion in the month of June.

Anar Butterfly

Scientific Name: *Virachola isocrates* Fabr.

Order : Lepidoptera

Family : Lycaenidae

Host Plants: Pomegranate, guava, loquat, orange and tamarind.

Distribution: Throughout India.

Nature of Damage: The anar butterfly is an important pest of pomegranate and causes heavy damage to its cultivation and causes heavy damage to its cultivation. The larva is a damaging stage, which bores into the fruit and feed on its contents. It creates a lot of mess and offensive smelling matter oozes out from the enterance hole. The wound caused by this pest gets infected with a number of bacteria and fungi thereby making the fruits rot. The holes made in the fruits by larvae are usually visible. During severe infestation the loss caused by the pest is estimated to the tune of 90%.

Life History: There are four developmental stages are found in its life cycle which have been described below:

Egg: The female butterfly lays whitish, shiny egg singly on the lower ends of the flowers, on the fruit surface and sometimes even on the leaves near the fruits. The hatching period of eggs varies from 7-10 days.

Larva: The young caterpillar instead of feeding on the leaves bores into the fruit and feed inside. It first feeds on the pulp and then to the seeds. When the larva is full grown, it comes out from the hole bores through the hard shell of the fruit and spins of web by which it at taches the fruit secured to the stem. This is done to prevent the fruit from falling down. The caterpillars are short, dark brown in colour with hairs and whitish patches on the body. The larval period lasts for 18-47 days. A single fruit may have about half a dozen caterpillars.

Pupa: The full grown caterpillars pupate either in the damaged fruits or on the stems holding them. The pupa is dark brown in colour. The pupal period varies from 7-34 days depending upon environmental conditions.

Adult: The butterfly is about 2.5 cm across the wings and glossy violet in colour in case of male and violet brown in case of female. The hind wings possess a tail like prolongation. The eyes have a rim of white scales.

The life cycle is completed in 33-93 days and breeding of the pest continues throughout the year with varying speeds depending on environmental conditions. There are 3-4 generations in a year.

Control

1. Bag the young fruits with stiff paper or polythene or coarse paper.
2. Spraying the trees, repeatedly with 0.07% endosulfan or 0.15%, carbaryl or 0.03% phosphamidon.
3. Destroy the infested fruits to minimise the damage.

Ber Fruit Fly

Scientific Name: *Carpomyia visuviana* Costa

Order : Diptera

Family : Trypetidae

Host plant: Ber.

Distribution: The pest is distributed in all parts of India, West Pakistan and the Middle East extending upto Southern Europe.

Nature of Damage: The damage is caused by the maggots of the fly which feed on the pulp inside the fruits. The fruits attacked by the pest become unfit for human consumption. During severe infestation usually in the month of February, it is difficult to get even a single sound and healthy fruit from hundreds of trees. During the peak period of attack the number of maggots is more than one inside the fruit and the whole pulp is consumed and decay set in by the maggots with the result that when damage has been completed, nothing but a semi-solid decayed dark brownish mass which smells offensively, is found within the fruit.

Life History: The four developmental stages namely egg, larva, pupa and adult are found in its life cycle, which have been described below:

Egg: The eggs are laid by the female fly singly or in clusters of 4-5 under the skin of the fruit and from the punctures thus made oozes out a little fluid which often dries up in the form of globule or dot of resinous material on the fruit. The hatching period varies from 4-5 days.

Larva: The young maggots after hatching from the eggs bore into the pulp and feed. The full grown maggot is amphineustic, creamy white pointed anteriorly and ends in a broad truncate last segment. The dorsal vessel is prominent. The full grown larva measures '6 mm in length and 1.5 mm in breadth. The maggots moult two times in its life time and larval period lasts for 9-12 days in March and 22 days in November-December.

Pupa: The maggots before pupation greatly contracts and thickens and pupates in the 1st larval skin which forms the puparium. Pupation generally takes place in the soil at a depth of 4-6 cm. usually the maggots leave the fruit for pupation out side the fruit but sometimes the drying state of the fruit may prompt the maggots to pupate within the fruit also. The pupal period lasts from 11-31 days during March-April and 45 to 87 days in winter. The well developed pupa measures 4.5 mm in length and 1.80 mm in breadth and light brown in colour.

Adult: The head of the freshly emerged fly is light almond brown. The wings are hyaline and transparent. The fly is about the size of a housefly, slightly more slender, with a pale, brown body, with an oval whorl of brownish black spots behind the head.

The pest is active in autumn and spring and the summer and winter are passed in the pupal stage. The total life cycle is completed in 25 to 117 days and there are 3 generations in a year.

Control

1. The attacked fruits should be collected and burnt or fed to the cattle.
2. Spray the trees with 0.07% endosulfan or 0.15% carbaryl before the fruits mature at intervals of 15 days.
3. Wild ber bushes should not be allowed to grow near the ber plantations.

Citrus Leaf Miner

Scientific Name: *Phyllocnistris citrella* Staint.

Order : Lepidoptera

Family : Gracillariidae

Host Plants: Citrus, pomelo, willow, cinnamon and *Loranthus* spp.

Distribution: It is found in all parts of India and occurs in many countries of South East Asia and Australia It is known to be a serious pest of citrus nurseries in Tamil Nadu, Madhya Pradesh, Assam, Uttar Pradesh and Punjab.

Nature of Damage: The caterpillar causes damage by making zig-zag silvery mines in the young leaves. Due to the extensive mining by the pest, the leaf suffers badly, gets deformed and irregularly curled up in shape, unhealthy in look and defective in its function and finally it dries and falls off. On the older leaves, brownish patches are formed which serve as foci of infection for citrus canker.

The succulent leaves with thin epidermis are more suitable for the penetrating and mining activities of the tiny larvae. The larvae feed more on plant sap than on the tissues. As a result, the leaves turn pale, curl up badly and dry off. In case of the nursery plants the entire stock may be ruined. In case of larger trees, photosynthesis is adversely affected due to which an appreciable reduction occurs in yield.

Life History: The four developmental stages namely egg, larva, pupa and adult are found in its life cycle, which are described below:

Egg: The eggs are usually laid singly on the underside of the leaves and tender shoots. They are mostly found 2 to 3 in each leaf. The eggs are minute, flattened, transparent and mostly laid near the mid rib. The hatching period varies from 2-10 days depending upon the environmental conditions.

Larva: The newly hatched larva is legless which enters the leaf tissue and begins to feed inside it and mines the leaf lamina. They form galleries within which they remain confined for the rest of their immature life. The full grown larva measures 5.1 mm in length and is pale yellow or pale green with light brown mandibles which are well developed. The larval period lasts 5-30 days.

Pupa: The full grown larva comes-out and pupates near the margin of the leaf which folds upto provide a sort of cover over the pupa. In the form of its body, the pupa is slightly brownish. It is partly exposed through the gallery wall and has a spine on its head with the help of which it pierces through the wall when it emerges as a moth. The pupal period varies from 5-25 days and the moths are commonly seen resting on the trunks of the trees near the ground.

Adult: The adult is a tiny greyish moth with a wing span of 8 to 10 mm. The forewings are white with two narrow grey stripes and the hind wing has a pale grey fringe.

The life cycle is completed in 14-68 days depending on the environment. The cycle of generations continues practically the whole year except that the development gets prolonged in the colder months. The moth activity may cease if it is very cold during December and January.

Control

1. Collecting and burning the mined leaves found useful.
2. The orchard should be kept free of wild citrus hedges, which serve as breeding places of the pest.
3. Spraying the trees with anyone of the following insecticides at 15 days intervals:
   - *(i)* Methyl demeton : 0.03%
   - *(ii)* Phosphamidon : 0.03%
   - *(iii)* Dimethoate : 0.03%
   - *(iv)* Monocrotophos : 0.04%

Citrus Psylla

Scientific Name: *Diaphorina citri* Kuwayanae.

Order : Hemiptera

Family : Psyllidae

Host Plants: All species of citrus and plants of the family Rutaceae.

Distribution: It is found in several parts of India and in the neighbouring countries, *viz.*, China, Formosa, Japan, Mayanmar, Sri Lanka, the East Indies and New Guinea. In India it is most destructive in Punjab, Himachal Pradesh and Maharashtra.

Nature of Damage: Damage is caused by the nymphs and adults. They suck the sap of the fresh tender parts of the host. The vitality of the plants deteriorates and the young leaves and twigs stop growing further. The leaf and flower buds may wilt and die. The small fruits formed in the spring fall off prematurely. The honey-dew excreted by the nymphs acts as a focus for the growth of the moulds on the leaves which adversely affects the photosynthesis. It is also thought that the insect produces a toxic substance in the plants with the result the fruits remain undersized, poor in juice and are insipid. The pest is also responsible for spreading the greening virus. If the pest is not controlled in time, the entire orchard may be lost.

Life History: There are three developmental stages namely egg, nymph and adult found in its life cycle, which are described below:

Egg: The adults start breeding in February-March and lay on an average 500 almond-shaped, orange and stalked eggs on tender leaves and shoots of citrus trees. The lower end of egg stalk is embedded in the plant tissue. They are laid either singly or group of two or three which are arranged in straight lines. There may be as many as 50 eggs in a place. The hatching period varies 4-6 days in summer and 10-20 days in winter.

Nymph: The newly hatched nymphs are light yellow in colour and have a tendency to stick close to the egg-shell. The nymphs are flat, louse-like and are seen congregated in large numbers on the young leaves and the buds, but in the later stage they may migrate to the older leaves. There are five nymphal stages and the development is completed in 12-25 days which may extend 34-36 days in December-January. When the nymphs are full grown, they migrate to the lower surface of the leaves where they change into adults.

Adult: The adult is small, 3 mm in length, active in habits, and rests on the leaf surface with closed wings. The insect is brown with its head lighter brown and pointed. The wings are membranous, semi-transparent, with a brown band in the apical half of the fore wings. The hind wings are shorter and thinner than the fore wings. The females live longer than the males.

The adults copulate 4-8 days after emergence and the females start egg laying immediately. The life cycle is completed in 20-64 days depending upon the environmental conditions. There are 8-9 overlapping generations in a year. During severe winter only adults survive, similarly in hot period of May-June, the adults are predominant, although other stages may also be found.

Control: The pest may be controlled by spraying the trees with any one of the following insecticides:

(i) Phosphamidon: 0.03%

(ii) Methyl demeton: 0.05%

(iii) Dimethoate: 0.03%

(iv) Malathion: 0.05%

(v) Monocrotophos: 0.04%

Lemon Butterfly

Common Name: Neebu ki surhi, Neebu ki titli.

Scientific Name: *Papilio demoleus* Linn.

Order : Lepidoptera

Family : Papilionidae

Host Plant: Citrus.

Distribution: It is widely distributed in the Indian subcontinent and is found right from Arabia in the West, to Formosa in the East.

Nature of Damage: The larva of the butterfly is damaging stage which feeds on the leaves of citrus plants of all kinds. The caterpillar cuts the leaves irregularly in the nursery. The leaves of young plants are completely eaten up. Thus the young plants suffer more by the attack of this pest. In severe infestation the larvae may eat even tender shoots. The pest is generally active from March to November and severe outbreaks generally occur during August-September.

Life History: The following four developmental stages are found in its life history:

Egg: The female butterfly lays yellowish-white eggs on young leaves and tender shoots. The eggs are smooth round and are generally laid scattered and singly. Sometimes they may be found in group of 2 to 5. A female can lay 75-120 eggs in her life span. The hatching period of eggs varied form 3-4 days in summer and 5-8 days in winter.

Larva: The newly hatched larva is about 3 mm long and initially feeds on the egg-shell and this habit persists through out its life. Larva eats exuvae after each moult. The young caterpillars are brown with white irregular markings in the body surface and they look like as if they are not caterpillars but some irregular masses of bird excreta. Obviously, this is an adaptation to escape the notice of their enemies like birds who would not like to pick-up their own droppings. The larvae become full grown in 8-16 days during summer and in about 28 days during winter. The full grown larva is yellowish green in colour and measuring 40 mm in length and 6.5 mm in breadth. It possesses a horn like structure on dorsal side of last abdomninal segment.

Pupa: The pupation takes place on the leaves or stem with a silken thread secreted by it. The pupa is husky in colour and passed winter in this stage. The pupal period varied 7 days in summer and 56 to 98 days during winter.

Adult: The butterfly is large and conspicuous with black and lemon yellow marks on its wings and with two blue eye like marks on the hind wings in the anal region. It is about 28 mm long and has black clavate type of antennae.

The butterfly after one day of emergence copulates and starts egg laying on 2nd or 3rd day.

The life cycle is completed in 20-117 days depending on environmental conditions and many generations are found in a year.

Control

1. During severe infestation the caterpillars may be picked and destroyed.
2. Any one of the following insecticides may be sprayed on a tree after mixing in 15 lit. of water.
   (i) Endosulfan 35 EC — 25-30 ml.
   (ii) Phosphamidon 85% — 5-6 ml.
   (iii) Formothion 25 EC — 30 ml.

Natural Enemies: *Ericia pimphalidophaga* is a larval parasite.

Singhara Beetle

Scientific Name: *Galerucella brimanica* (Jacoby)

Order : Coleoptera

Family : Chrysomelidae

Food Plant: Singhara.

Distribution: It is widely distributed in Pakistan, Sri Lanka, Mayanmar and India.

Nature of Damage: Damage is caused by both grubs and beetles which feed on the leaves of waternut floating on the surface of the water. The attacked leaves are riddled with holes and rendered useless, as a result of which the growth of the plant is checked and the formation of flowers and the production of fruits greatly reduced. Sometimes the integument of *singhara* fruits at the surface of the water is also damaged which results in the rotting of the fruits. A badly damaged crop turns black instead of looking green maximum damage is caused by the pest during August and September.

Life History: The four developmental stages namely egg, larva, pupa and adult are found in its life cycle which have been described as below:

Egg: The beetle lays eggs on the upper surface of the leaves in clusters of 10-20. The eggs are rounded and reddish-brown in colour. One female may lay as many as 300 eggs during its life span. The eggs are securely glued on to the leaves. The hatching period of eggs varies from 6-9 days.

Larva: The newly hatched grubs start feeding on the upper epidermis of the leaves. The young grub is light brown in colour and undergoes in moultings. The full grown larva is dark brownish-grey in colour and about 6-7 mm long. The larval stage lasts for 10-22 days.

Pupa: Before, pupation, the grub becomes sluggish and enters the pre-pupal stage lasting for 1-2 days. The pupation takes place on the leaf surface and like the eggs the pupa are also firmly flued to the leaves. The pupa is oval shaped and orange red in colour with dark brown patches. The period lasts 3-7 days after which the adult beetle emerges out.

Adult: The adult beetle is blackish-brown in colour, 6 mm in length and 3 mm in breadth. It possesses a large hump in the middle of the body and eyes are black. The adult beetles live for a month or so, although some of them may live for three months.

The pest is active generally throughout the year but in places where water is receded, it aestivates in cracks and crevices in the sides of ponds. These beetles again become active in July. The total life cycle is completed in 21-41 days and there are generally 6-7 generations in a year.

Control

1. Collect egg masses and beetles from the leaves and destroy them in the early stage of the crop.

2. Dusting the crop with any one of the following insecticides per hectare:
   (i) Malathion 5% dust : 25 kg.
   (ii) Carbaryl 10% dust : 25 kg.
   (iii) Carbaryl 5% dust : 30-35 kg.

Fruit Sucking Moths

There are about 20 species of fruit sucking moths out of which following are more important

Ophideres conjuncta (fullonia) Cramer

O. matema Cramer

Achoea Janata Linn.

Calpe emarginata Fabr.

Common Name: Neebu ke phalon ka patinga

Scientific Name: *Ophideres* spp.

Order : Lepidoptera

Family : Noctuidae

Host Plants: Citrus, grape-fruit, pomegranate, malta and orange. The caterpillars also feed on the wild plants.

Nature of Damage: Generally moths and butterflies damage the plant or plant products only during their larval stages but fruit sucking moths as their common name indicates damage fruits in their adult stage. The moths damage the outer skin and then suck the juice of fruits. Usually ripening fruits are attacked but unripe fruits are not always immune from such attacks. Severely attacked fruits rot and fall to the ground. The damage also exposes the fruits to be attacked by fungi and bacteria etc. The damage caused by them is estimated to the tune of 20-40 per cent.

Life History: The four developmental stages *viz.*, egg, larva pupa and adult are found in its life cycle which are described as under:

Egg: The female lays eggs singly on the underside of the leaves of creepers namely gurch *(Tinospora codolia)* and vasnable *(Cocculus villosus).* The eggs are round or oval measuring 1 mm in diameter and shining pale-green in colour. A female lays 200-300 eggs in her life span. The eggs hatch in 8-10 days.

Larva: The larvae start feeding on the host plants (wild creeper and plants) within 24 hours of hatching. The caterpillars undergo 5 moults and become full grown in 12-21 days. The full grown caterpillar is stout, hump backed (semi looper) and measures SO to 60 mm in length. The larva is velvety-blue in colour with dorsal and lateral sides covered with bright-red and yellow spots.

Pupa: It pupates in a leaf-fold under a very thin whitish-pale patchment made up of silken threads. The pupa is stout and reddish-brown in colour. Pupal period lasts from 8 to 15 days.

Adult: The adult moth is 2.5 cm long with a wing span of 9 cm. The fore wings are brownish green while hind wings are orange or yellow in colour with black spots in case of *O. conjuncta.*

The moths of *O. matema* are light orange to brown in colour. Fore wings are dark brown while hind wings are orange-red in colour with two black spots.

The fore wings of *O. ancilla* are yellowish-green with dark markings and white spots.

The life cycle is completed in 31-50 days and generally 3 generations are found a year. The moths are nocturnal in habit and remain active during the rainy season, mainly from July to September.

Control

1. The wild creepers, plants which are the food plants of larvae should not be allowed to grow near the orchards.
2. The moths may be attracted in light traps and destroyed.

3. Smoking in the orchards escape the odour of ripening fruits which attract the moths.
4. The fruits may be bagged with plastic or cloth bags.
5. Bait the moths with lead arsenate and gur solution mixed in the proportion of 1: 160 and hang on the branches in small pots. Few drops of vinegar may be mixed to attract the moth.

Castor Semilooper

Scientific Name: *Achoea janata* Linn.

Order : Lepidoptera

Family : Noctuidae

Host Plants: Adult moths suck the sap of ripening fruits while larva feeds on mainly castor leaves.

Distribution: It is distributed through out the Indian continent.

Nature of Damage: The moths also damage castor plants and guava fruits besides citrus fruits. The moths pierce the fruits and suck the juice from them. One fruit may have as many as 20 to 25 punctures in it as a result of which it rots and falls to the ground. The infested fruits are also attacked by fungi and bacteria etc. which initiate early roting, the larva also causes heavy loss to castor plant during severe infestation the entire leaves are eaten up leaving veins and stalks only.

Life History: There are four (developmental stages namely egg, larva, pupa and adult in its life cycle described below:

Egg: The female moth lays eggs scattered on the under surface of leaves. The egg is comparatively large (0.9 mm long) and has ridges and furrows on the surface radiating from a circular depression at the apex. It is blue-green in colour and difficult to spot out on the leaf. A female lays about 400-500 eggs in her life time. Hatching-period lasts from 2-5 days.

Larva: The larva emerges by biting a hole to the egg-shell which forms the first meal of the newly hatched larva before it feeds on the leaf tissue. The first stage larva is about 3.5 mm in length. It moults 4 to 5 times and becomes full grown measuring 60-70 mm in length. The locomotory process involving the formation of a semi-loop of the larval body. On the dorsal side of the larval body white and red stripes are found. The larval period lasts from 5 to 20 days.

Pupa: Pupation takes place both in the soil and under fallen leaves and other rubbish at the edge of the field. The larva prepares a loose cocoon of coarse silk to which particles of soil readily adhere. Some times pupation also takes place within the folded leaves on the plant itself. The pupal stage lasts from 10 to 15 days.

Adult: The moth is pale-reddish brown in colour with wing span of 60-70 mm. Both the fore and hind wings have broad zig-zag decorative markings and large pale dark brown patches. They are nocturnal in habit and attracted in light.

The winter is passed in pupal stage and moths emerge out in late February or in early March. The female starts egg laying after 3-4 days of copulation. The life cycle is completed in 30-44 days, there are 5 to 6 generations in a year.

Control: The moths may be controlled in a similar way as mentioned in the case of *Ophideres* spp. The larvae attacking castor crop may be controlled as follows:

1. Dusting the crop with 5 or 10% BHC dust @ 25 or 30 kg/hect. found effective.
2. The spraying of the crop with any one of the following insecticides has also been found useful:

   (a) Endosulfan 35 EC - 1.25 lit/hect.

   (b) Phosphamidon 85% - 225 ml per hect.

   (c) Methyl parathion 50 EC - 1 lit/hect.

Potato Tuber Moth

Common Name: Alu-ki-surhi

Scientific Name: *Pthorimoea (Gnorimoschema)operculella* Zeller.

Order : Lepidoptera

Family : Gelechiidae

Host Plants: Potato, tomato, brinjal and tobacco etc.

Distribution: This is found in India, China, Australia, New Zealand and Hawaii. In India it is distributed every where except the hilly tracts of the country.

Nature of Damage: The damage is caused by larvae which bore in to the tuber and feed on the pulp. Damage is done in the field as well as in godowns but it is a serious pest of stored potato. In the field it causes damage from October to February-March. Initially the caterpillars mine the leaf bore in to branches and stem and feed inside but during severe infestation they may eat leaves also. The exposed tubers in the field are also subjected to their attack. With these infested potatoes they reach to godown and multiply. The damaged tubers are unfit for seed as well as for human consumption. In addition, the attacked tubers are exposed to bacterial infection which leads to rotting. The infested tubers may be differentiated by the presence of black excreta near the eye bud. Storing of potatoes in cold storage reduces the tuber moth attack almost completely.

Life History: The four development stages, *viz.*, egg, larva, pupa and adult are found in its life history which are described as under:

Egg: The female moth lays eggs in or around eyes or cracks of skin of the tubers in godowns. In the field eggs are laid on the lower surface of the leaves. The eggs are pearly-white in colour when freshly laid, turning brownish or blakish lateron and measuring 0.6 mm in length and 0.4 mm in breadth. A female lays 100 to 150 eggs in her life span. The eggs each in 3 to 4 days.

Larva: The newly hatched larva is about 1.5 mm long white in colour with black head.

It bores in to the tuber and enters in it through a minute hole. The caterpillar makes tunnels in all directions in the pulp of the tuber and becomes full grown in 10-20 days. It moults 5 times and full grown caterpillar is pinkish in colour and measures about 22 mm in length. A single potato may have several caterpillars. The full grown larva comes out of the tuber for pupation.

Pupa: The larva pupates among the potatoes or crevices of their containers or on fallen leaves in the field. It forms a dirty cocoon of 6.5 mm long for pupation. The pupa is brown and measures 6.5 mm in length and 2.5 mm in breadth. Pupal stage lasts from 7 to 10 days.

Adult: The adult moth is greyish-brown in colour measures 6 mm long with a wing span of 12 mm. The fore wings are greyish-brown with yellow longitudinal bands and hind wings have a fringe of hair on the posterior margin. The adults are nocturnal in habit and attracted in the light.

The mating of moths takes place soon after their emergence and female starts egg laying after 1 or 2 days of copulation. The life cycle is completed in 21-36 days and 6 to 8 generations are found in a year.

Control

1. Affected tubers should be picked and rejected before storage.
2. The potatoes are stored in sand (2.5 mm thick) so that the female could not lay eggs on them.
3. In the field when infestation starts, crop may be sprayed with endosulfan 35 EC @ 1 lit/hect. in 750 lit of water.
4. If the potatoes are stored for seed purpose the malathion dust may be mixed @ 1.5 kg/m tonnes of tubers.
5. Fumigation with aluminium phosphide @ 7 tablets/28 cubic meter space is effective in killing the caterpillars.

Natural Enemies: The following are the effective larval parasites:

(i) Bracon hebitor

(ii) Serratia marcescens

(iii) Microbracon gelechiae

Rice Weevil

Common Name: Ghun, Killa, Ghanera, Sund Wali Surasari.

Scientific Name

(i) *Sitophilus oryzae* Linn.

(ii) *Sitophilus granarius* Linn.

Order : Coleoptera

Family : Curculionidae

Food: Rice, paddy, wheat, barley, maize, jowar etc.

Distribution: It is cosmopolitan in distribution but is generally much more injurious in warm humid countries. India is considered to be its original place.

Nature of Damage: This is the most destructive insect pest of stored cereals in the world. It is called the rice-weevil only because it was found to be infesting rice when it was described for the first time. Both, adults and the larvae (grubs) damage the grain on which they feed voraciously so mesh so the grain is rendered unfit for human consumption as well as for seed purposes. The grub is more injurious than adult it makes a tiny hole in to the grain, enter inside and feed on the content of the grain leaving only shell.

Usually it is active throughout the year but serious damage is caused from July to November. Moisture is important factor for its development. Under favourable conditions, it is found in great number and due to respiration heat spots are found in the grains. In cases of heavy infestation the grain becomes a mass of broken vegetable matter. Some times black fungus also grows.

Life History: The four developmental stages namely egg, larva, pupa and adult are found in its life cycle which have been described as below:

Egg: Eggs are deposited by the female after 5 days of adult formation. The mother-weevil makes a small cavity in the grain by means of its powerful jaws and then lays an egg in the cavity and cover it with a gelatinous fluid. Usually one egg is laid in a grain but some times 2 or 3 eggs may be deposited. The eggs are white in colour, oval in out line and measures about 0.7 mm long and 0.3 mm broad. A single female can lay 250-300 eggs in her life span. The eggs hatch in about two days during summer and 6-9 days during winter.

Larva: The tiny grub immediately after hatching bore into the grain kernel. The grub is white in colour with yellowish brown head legless and provided with biting type mouth parts. It remains inside the grain feeds on starchy content and emerges on reaching the adult weevil stage. It moults 3 times inside the grain and becomes full grown in 25-35 days. The full grown grub is yellowish-white in colour and about 0.3 cm in length.

Pupa: The grub makes a pupal shell inside the grain and pupates therein after passing a day or two as pre-pupal period. The pupa is at first dirtywhite in colour but later on it changes to dark brown and measures 2.5 mm to 3.0 mm in length. The pupal stage lasts for 4-7 days during summer and 14-20 days during winter.

Adult: It is a reddish-brown weevil which turns into black brown, about 3.5 mm in length and its head has a slender pointed forward projection (snout) with a pair of stout mandibular jaws at its extremity. The snout is about 1 mm long possess clavate antennae at its base. The adult of *S. oryzae* are light brown in colour having 2 pairs of spots on each elytra while these are absent in S. *granarius.* The S. *granarius* can not fly as its wings are fused while *S. oryzae* is strong flier.

The life cycle is completed in 36-69 days and 6-8 generations are found in a year. The adult weevil may live 21-28 days in summer and 3 months in winter. The weevil hibernates during severe winter in cracks and crevices.

Its life history is mainly dependent on humidity and temperature. At 70% humidity and 17°C temperature it is completed in 220 days, as the humidity increases it takes much less time to complete the life cycle. During rainy season, when conditions are favourable a pair of weevil may produce 10,00,000 insects in a season.

Control: Control will be given in last for all the storage pests.

Natural Enemies: The following are the important natural enemies:

Egg Parasite: *Acarapsis doctamite*

Grub Parasite: Larrophagous distingnendus forst

Aplastomorpha calanrae how

Pendiculeides ventricosis new port-mite.

Khapra Beetle

Common Name: Khapia, Pat, Wantari.

Scientific Name: *Trogoderma granarium* Everts

Order : Coleoptera

Family : Dermestidae

Food: Wheat, Rice, Jowar, Bajra and Maize etc.

Distribution: It is mainly distributed in India, Africa, Mayanmar and China and also found in countries where temperature ranges from 32-44°C.

Nature of Damage: Unlike other beetles the adult in this case is harmless and the damage is caused only by the larval stage. The grub first attacks at embryo point but later when infestation is severe other parts of the grain are also badly damaged. Usually the infestation occurs in superficial layers of the grains, as the pest do not penetrate beyond some depth. This is due to the fact that in its egg pupal and adult stages the beetle is highly susceptible to reduction in oxygen tension.

Maximum damage by the pest is caused during July to October. Since it mainly feeds on embryo, germination power of infested grain is adversely affected.

Life History: There are four developmental stages, *viz.*, egg, larva, pupa and adult in its life cycle, which are described as under:

Egg: The female starts egg laying after 5-6 days of copulation. The eggs are generally laid singly or rarely 2-5 together among the grain. An individual egg is round small, semi-transparent white in colour and measures 1/2 mm long. The eggs are laid sometimes as many as 26 per day and a single female can lay about 125-150 eggs in her life time. The eggs hatch in 6-16 days depending upon the environmental conditions.

Larva: The newly hatched grub is yellowish white in colour but later turn brownish with transverse yellowish-brown bands and bundles of reddish-brown hairs over and along the body. The long hairs are also found at the hind end of the body. The male grub takes 19-28 days in development while female grub 20-37 days. Under adverse conditions, they may prolong their life to as long as 4 years, living for the greater part of it even without food. Usually they moult 4 times but under adverse conditions they moult 10-12 times. It passes winter in its grub stage and remains hidden in cracks and crevices. Full grown grub is about 5 mm long.

Pupa: Pupation takes place in the last larval skin among the grains or in cocoons. The pupa is dark brown in colour and having reddish-brown hairs on it. The pupal stage lasts for 4-27 days depending upon environmental conditions.

Adult: The female beetle is a small rather oval, pale red-brown or black in colour and measures 2.5 mm in length and 2.00 mm in breadth. It has distinct markings on the wing covers. The male is almost half the size of the female. The antennae are small and capitate type.

The adults are ready to start the next generation very soon after emergence. The life cycle is completed under favourable conditions in 32-84 days. There are generally 20-22 generations in a year. The adult beetle may survive 25-35 days.

Natural Enemies

(i) Reduvid bugs predator as grub stage

(ii) *Grub parasites: Anisopteromalus calandrae,* Harward, *Dinarmus laticaps, Ashmead-Synopeas'* sp.

Lesser Grain Borer

Common Name: Lal Surhi, thuthun heen surhi

Scientific Name: *Rhizopertha dominica* Fabr.

Order : Coleoptera

Family : Bostrychidae

Food: Wheat, rice, maize, jowar, barley, gram, flour and dried fruits etc.

Distribution: It is cosmopolitan in nature and mostly found in India, Pakistan, America, Argentina and Australia.

Nature of Damage: The damage is caused by both beetle and grub but beetle is more harmful than the grub. In some places it is considered to be next to rice weevil but other places it is reported to be causing more damage than even rice weevil. The grubs eat out the starchy contents of the grains leaving the outer husk only while beetles destroy whole grains which are reduced to frass and waste flour. The first stage larva is straight and can bore into sound grain but in the later stages of growth it is curved, therefore, it becomes difficult to them to penetrate the grain. The grubs which are unable to penetrate the grain feed on the waste flour left by the adults. Some times adult beetles attack the grain in the field it self. The serious damage is caused during May to August

Life History: The four developmental stages, *viz.,* egg, larva, pupa and adult are found in its life cycle which are described as below:

Egg: The beetle starts egg laying often 4-5 days of becoming adult. The eggs are laid from April to November. The female generally lay eggs on the grain near the embryo-end, which is comparatively soft and from where the newly hatched grub easily enter the grain.

The eggs are pear shaped and are laid either singly or some times in clusters. They are initially white in colour which later turns into pinkish. A female can lay 300-500 eggs in her life time. The hatching period varies 5-6 days during summer and 15-21 days during winter.

Larva: The newly hatched grub is quite active and white in colour, soon enters into grain generally near the comparatively soft embryo end and passes the rest of its life therein. Some of them remain feeding on the powdered starchy material without entering the grain. The larva undergoes 4-5 moults. The full grown larva is dirty white in colour with a light brown head and curved abdomen, some what 'C shaped. It measures 2-3 mm in length and covered with tiny hairs. The grub period lasts for 40-50 days.

Pupa: Pupation takes place either inside the grain or out side the grain depending upon where grub lives. The pupa is dirty white in colour and irregular in shape. The pupal period is of 7-8 days.

Adult: It is a small polished, dark brown or black beetle measuring 3-4 mm in length and a slender cylindrical body with its head turned down under the thorax. The antennae are 10 segmented and terminate in a prominent club.

The life cycle is completed in 56-84 days and 5-7 generations are found in a year. The insect hibernates both as a grub and adult during December to February. They resume activity in March and adults starts to lay eggs in April after 4-5 days of copulation.

Natural Enemies

(i) Lariophagus distinguendus Frost-grub parasite

(ii) Pediculoides sp.—A mite predating on eggs and grubs.

Red Rust Flour Beetle

Common Name: Sursali, Lalsuri.

Scientific Name: *Tribolium castaneum* Herbst.

Order : Coleoptera

Family : Tenebrionidae

Food: Stored grains and stored products of various kinds such as flour, maida, suji, cotton, ground nuts, beans and dry fruits etc.

Distribution: It is found all over the world and in India the pest is mainly distributed in Uttar Pradesh, Bihar, Madhya Pradesh, Punjab, Haryana and Rajasthan.

Nature of Damage: The insect does not cause damage to whole grain but mainly feeds on broken or attacked grains of other insects. It is generally found in association with Khapra beetle, rice weevil and lesser grain borer and does not cause much damage as done by these pests. The insect is specially harmful to flour, maida and suji. In cases of severe infestation flour and flour products turn greyish yellow and become mouldy which produces pungent smell and become unfit for human consumption. Both grub and adult are responsible for causing damage. The maximum damage is caused during rainy season.

Life History: The four developmental stages, *viz.*, egg, larva, pupa and adult are found in its life cycle which are described as below:

Egg: The female starts egg laying after 2-3 days of copulation. The eggs are generally laid singly in flour or among grains and being sticky, covered with flour or dust and are difficult to detect. A single female can lay 400-550 eggs in her life time which are cylindrical and white in colour. The eggs hatch in 4-12 days depending upon environmental conditions.

Larva: The young grub is small, slender and cylindrical in appearance with its body segments having a number of fine hairs and the terminal segment being in addition provided with a pair of spine like appendages. At the time of hatching the grub is 1 mm long which grows up to 5 mm when fully developed. It moults 6-7 times and becomes full grown in 27 to 90 days depending upon the food available and prevailing temperature. At 30°C temperature it becomes full grown in 22 to 25 days.

Pupa: Usually, pupation takes place on the surface of the food. Initially the pupa is white in colour which gradually turns into yellowish and covered with hairs on the dorsal surface. Pupal stage lasts for 5-9 days.

Adult: It is small reddish-brown beetle measuring 3.5 mm in length and 1.2 mm in breadth. The head, thorax and abdomen are distinct. The antennae are well developed and last three segments form an enlarged terminal end. The life cycle is completed in 38-114 days and many generations are found in a year. The beetle hibernates during winter and resumes activity in spring when they copulate after 1 or 2 days of their emergence.

Control: Like other storage pests.

Pulse-beetle

Common Name: Dhora, Chiraiya, Dhorna, Dhanur.

Scientific Name: Following 3 species are found in India:

(i) *Callosobruchus chinensis* (Linn.)

(ii) C. *maculatus* (Fab.)

(iii) C. *analis* (Fab.)

Order : Coleoptera

Family : Bruchidae

Food: Arhar, mung, urd, pea, gram, lentil, cowpea, soyabean, moth and beans etc.

Distribution: It is world-wide in distribution and has been found in all parts of India. It was first described from China in 1758 due to which the name chinensis was given to the species.

Nature of Damage: Generally damage starts in the leguminous pods in the field from where they are carried to storage godowns. However, they do really serious and spectacular damage under storage conditions. Both grubs and adult beetles are responsible for causing the damage. The grubs cause damage by eating out the entire content of the grain, leaving only the shell behind. The damaged grains are unfit for human consumption as well as for sowing purposes. The maximum damage is caused during July to September. The losses caused by this pest to the pulses have been estimated to the tune of 40 to 50 per cent in storage.

Life History: Four developmental stages, *viz.*, egg, larva, pupa and adult are found in its life cycle which are described as below:

In the field, the eggs are laid on the developing pods either on the outside or inside the pods. The tiny grubs after hatching bore the grains and develop inside. The entry hole is generally too small and gets healed up soon. Under storage conditions, the eggs are glued on the grains, singly but many eggs may be seen on a single grain. The eggs, when freshly laid, are translucent, smooth and shining, which later become yellowish-white and very prominent against the contrasting back ground of the grain surface. A single female lays about 60-90 eggs in her life time which hatch in 4 to 5 days.

Larva: The young grubs bore into the pod or grain and develop inside where they grown up. The full grown grub is about 5 mm long cylindrical in shape, fleshy, wrinkled and white in colour except head which is brown, in colour. It has three pairs of legs and each leg possesses a claw at the terminal end. The grub moults 4 times and becomes full grown within 2-3 weeks.

Pupa: When grub is full grown it cuts a circular disc in the seed coat which just remains in position and it pupates inside the grain. The pupa is brown in colour and pupal period lasts for 4 days during favourable conditions, which extends to 4 weeks during winter. After which the beetle emerges by further cutting out the circular disc into an emergence hole.

Adult: The adult is a small chocolate coloured beetle measuring 3.2 mm in length. The head is small with blunt snout, the femur of the hind leg specially thickened, truncate elytra which do not cover the posterior portion of the abdomen. The antennae are serrate type in case of female and pectinate in case of male.

The pest hibernates during winter as larva and adults begin to appear towards the end of March. The female starts egg laying after 3-4 days of copulation. The life cycle is completed in 25-58 days and 7 to 8 overlapping generations are found in a year. The adults are short lived and may not survive for more than two weeks.

Food: Wheat, maize, rice, barley, jowar etc.

Distribution: The pest is also known as Angoumois grain moth because it was first reported in 1736 from the Angoumois province of France. Now it is distributed through out the world including India.

Nature of Damage: The infestation of this pest starts when the crop is still in the field and the grain is in the milk stage. This infestation continues to increase unchecked right upto the time when the harvest is waiting for threshing. The damage to the grain is always caused by the larvae which bore into grain. Once it has entered the grain, the larva eats out the kernel unseen. The infested grains are hollowed out by the larvae and filled with their excreta and webbing. Its infestation generally does not extend beyond a few inches below the surface of stored grain. Where the moisture content of the grain is high during storage, this pest is considered to be extremely serious. The maximum damage is caused during months of monsoon, *viz.*, July, August and September. According to *Pruthi* and *Singh* (1950) this pest causes more than 10% loss to various commodities.

Life History: The four developmental stages namely egg, larva, pupa and adult are found in its life cycle which have been described as below:

Egg: Female lays eggs in depressions, cracks and crevices in floors or holes in the grains, The eggs are oval in shape, about 0.5 mm in length, with both ends rounded and the surface finely scluptured. They are initially white in colour but soon change into bright red. A single female can lay upto 400 eggs which hatch in 4 days during summer and 7 days during winter.

Larva: The tiny larva crawls about a little and soon finds out a soft spot through which it enters the grain. When it is inside, closes the entry hole by a silken web and develops therein. The full grown larva is white with yellowish brown head and measures about 6 mm in length. The caterpillar moults 5 times and becomes full grown in 14-20 days which extends in winter. At this stage, the larva cuts out a circular exit hole, leaving over it just a sort of cap which can be pushed off easily by the future moth.

Pupa: Larva spins a silken cocoon inside the grain the transforms itself into a brown pupa. It measures 5 mm in length. The pupal period lasts for 7-10 days under favourable conditions.

Adult: The moth is small 6 mm in length, buff or yellowish-brown in colour with wing span of 12 mm. It has two pairs narrow pointed wings having long fringes. The head is small with black and possesses filiform antennae.

The female moth starts egg laying after 2 to 3 days of copulation. The life cycle is completed in 27-40 days depending on the environmental temperature and humidity. There are 4-5 generations in a year.

Control: Like other storage pest.

Natural Enemies: *Laemophloeus minutum* and *Psocids* are the predators on the larvae and pupae of the pest.

Prevention and Control of Stored Grain Pest

There is a problem before the Indian farmers to protect their grains during storage. It has been estimated by Bindra 1975 that about 25% damage is caused from different sources to the stored grain which comes to thousand million tons of food grains. Thus we see that to become self-sufficient in food and to improve the economic condition of the country it is utmost important to check the losses caused by various agencies during storage. Grain under storage conditions suffers damage mainly from rats, insects, mites and micro-organisms. The problem of rats basically different from others which needs special attention while rest three can be tackled together.

It has been proved by experiments that damage caused by insect, mites and micro-organism depends upon the following three factors:

(i) The moisture content of the grain.

(ii) Sources, availability and amount of oxygen in the storage.

(iii) The development of temperature gradient within the stored grain.

Moisture content: For the proper development each insect and mite species requires particular humidity which in stored grains depends on the moisture content of the grain. The minimum moisture content required by some storage pests is given below:

1. *Trogodeima granarium* 0 — 1.9%
2. *Rhizopertha dominica* 9.0 —10%
3. *Sitophilus oryzae* 9.5 — 11%
4. *Tribolium castaneum* 10%
5. *Corcyra cephalonica* 9%
6. *Cadra cautella* 10%

It is clear from the above that the moisture content of grains is mainly below 9 to 10 per cent, then infestation of most of the pests can be avoided except that of khapra beetle. The initial moisture content before storage may be brought down by sun drying or by grain driers. The grains will be fit for storage if they are cut into pieces by teeth. Care should be taken that dried grain should always be stored in moisture proof godown otherwise it will soon adjust its moisture in accordance with atmospheric humidity which will invite the infestation of insect pests.

Availability of Oxygen: Like humidity, every insect species requires a certain percentage of oxygen for its development and survival when the grains are stored in an air tight storage, the oxygen-content get reduced as it is consumed in respiration both by grain and by the insects, if present in the grain. Therefore, in the absence of minimum oxygen percentage, the development of an insect ceases and it dies. For example, khapra beetle required 16.8% oxygen content for the survival of eggs, therefore, its multiplication may be checked by lowering the oxygen content to this level.

Temperature: Temperature is the most important factor controlling the development and multiplication of insects. The insects survive and thrive within a certain limit of temperature beyond which their activity ceases. Certain insects form heat spots for their survival. The development of heat spot is due to the excessive heat produced either by insect multiplication or by microbial infection. Insect respiration produces both temperature and moisture increases, leading to favourable conditions for their multiplication. Therefore, efforts should be made to prevent the heat spots by way of cooling which may be done by forced aeration, turning of the grain and control of infestation by fumigation etc.

If these principles are kept in mind during storage the damage can be minimised upto a greater extent. Before discussing the preventive and control measures it would be worthy to know the sources of infestation during storage which are given below:

1. Some insect pests like rice weevil, pulse beetle

and grain moth, come with the grain from the field it self.

2. During threshing they may infest the grain and reach un-noticed to the godown.
3. If they are present in bullock-carts, trucks, rails and other vehicles may infest the grain during transportation and reach to the godown.
4. They may be present in the old bags and infest the healthy grain when stored on them.
5. They are generally hidden in the cracks and crevices of the walls in the godown and become active when grains are stored in such godowns.
6. The adults may reach to godown by flying and larvae by moving from the neighbouring places.

Preventive Measures

1. Godowns should be cleaned well and white washed, if possible, moisture proof godown should always be preferred.
2. All the cracks, crevices and holes present in the floor, walls and ceiling of the store should be filled up with cement and levelled.
3. The walls of the store may be painted with coal-tar from the ground up to the height of 1.5 meter.
4. The dirt, broken infested grains and sweeping of the stores should be removed and burnt before new grain is stored.
5. If the cracks, crevices and holes of the godown are suspected to harbour insects, the godown should be treated with any one of the following insecticides:
   (a) Fumigation with ED/CT mixture (Killoptera) for 24 hours @ 10 lit/40 cubic meter of space.
   (b) The ceiling and walls of godown may be sprayed with 0.5% malathion @ 3 lit/sq. meter.

Cleaning of Bags

1. Use of new bags as for as possible.
2. If the old bags are to be used, these should be disinfested by the following ways:
   (a) By dipping them in boiling water for about 15 minutes
   (b) By drying them in hot sun heat for about 6 hours.
   (c) By fumigating them with ED/CT mixture at the rate of 1 litre per 10 bags.
   (d) By dipping them in 1% malathion solution for 10 minutes.
3. About 20% of the room should be left free between the top layer of the bags and ceiling.

Cleaning of Grains and Precautions

1. Bullock carts, trucks, or other vehicles used for the transports of grains should be cleaned and washed preferably with phenyl water.
2. The grain should be thoroughly dried before storage, so that it does not have more than 9-10 per cent moisture.
3. The clean grain may be brought to the store direct from the threshing yard.
4. Only one kind of grain should be stored in a store as for possible.
5. If the grain is stored in bags, a layer of bhusa should be sprayed at the floor and bags should be kept 50 cm away from the walls.
6. Dunnage material (bhusa etc.) must be free from pests.
7. If the grain is already infested, it should be treated with ED/CT mixture @ 15 lit/40 q of grain before storing.
8. The grain may be stored with neem-seed kernel powder in the ratio of 100 : 1 to prevent the infestation of storage pests.

Remedial Measures: When the attack of stored grain insects is in progress in the godown, any one of the following fumigants may be used to check them:

1. ED/CT mixture @ 0.5 lit/metric tons of grain.
2. EDB ampule (3 ml) @ 1/q of grain.
3. Aluminium phosphide tablets @ I/metric ton of grain or 7 tablets per 28 cu. m. of space.

Storage Structures: Various storage structures of different capacity have been designed and recommended for storing the grain which prevent the infestation of rats, insects and micro-organisms. The most common are given below:

1. Pusa bin
2. Pant Nagar kuthla
3. Hapur bin

Pusa Bin: It is a rectangular structure constructed over a pucca brick floor and can be made as per requirement. A mud platform of roughly 1.7 m × 1.2 m × 7 cm (thick) size is made and a polythene film of 1.8 m × 1.4 m size is placed over it. A 7 cm layer of mud is then applied over the film. The inner wall (11cm thick) of structure covering the four sides is then constructed. The inner layer of the mud roof of the structure is then made by using 5 cm thick mud slabs prepared earlier and placing them over a wooden frame. An area of 5 cm × 50 cm size is however, kept open in one corner to serve as a main hole. The entire structure is then covered with the polythene film. Finally, the outer layers of the walls of 11 cm thickness are erected all round the structure covering the polythene film. A 5 cm thick mud plaster is also put over the polythene film at the top. For making the structure rat proof, the outer wall of the structure may be constructed using pucca bricks. When the structure is filled with grain the main hole is finally sealed with a square piece of polythene film.

Pant Nagar Kuthla: Pant nagar kuthla is simple modification of regular kuthla prepared by farmers in their house. To prepare Pant Nagar kuthla a plateform of 30 inches diameter and 2 inches thickness was made of the thick dough prepared from bhusa and rawtank mud. A layer of bitumin impregnated hessian cloth (Tar felt) was pasted on it. A 2 inches thick layer of mud wall of three feet height in continuation was constructed and left for drying. A tin plate was fitted the base giving 9 "projection around to avoid rat entry. A outlet hole of 6 inches diameter was left at one side. As soon as it was dried 'Tar felt' with outlet hole of 6 inches was paisted over this wall.

The tarfelt should always be in continuation of the previous layer and no gap should be left at any point. Over the Tarfelt layer, a layer of mud was plastered. Thus Pant Nagar kuthla of 60 inches height was completed till it reached to inlet hole of 12 inches diameter was also left. The lid for out-let and inlet were also constructed in the same way by sand witching 'Tar felt'. This kuthla was then dried till the relative humidity inside was 40-50 per cent.

Hapur-bins: There are various types of bins available for farmers use which may broadly categorised into two groups:

1. Indoor bins
2. Out-door bins.

Indoor bins: This group includes domestic designs of metal bins. Gharelu Thekka, Pucca kothi, welded wire mesh bin and RCC ring bin. Among these domestic designs are most popular.

Domestic Designs: In this group seven different types of domestic designs of metal bins have been developed, the capacities or which are ranging from 3 to 27.5 qtls. These are indoor bins and may be kept in a room or varandah under a roof. All these bins are fabricated using either 24 gauge of 22 gauge G.P. sheets of different standard sizes. The locking facility has been provided in all types of domestic bins. The inlets are provided at the top and different types of outlets are provided at the bottom of the bins to facilitate unloading the different commodities of food grains. These bins are found to be suitable for storage of wheat, pulses and seed grains.

Out-door Bins: It includes flat bottom metal bins, hopper bottom metal bins, composite bins and R.B. bins. Under this category flat bottom metal bins are more popular.

Flat Bottom Metal Bins: These bins are for out-door use and five different capacities ranging from 20 to 50 quintels are available in this design. The model type-2 bins are fabricated using 20-G. P. sheets and model type-3 bins fabricated using 18 gauge aluminium sheets. The bins are supplied to the users either in the form of loose components, semi -assembled or fully assembled depending upon the distance of the transportation.

In one design the bin can be erected on brick masonry base while in the other it can be erected on pre-fabricated elevated steel base. Since its height is not much, loading of the bin can be done manually through a simple lifting device provided. The bins are suitable for storage of wheat, paddy and maize. As aluminium is rust proof, periodical maintenance is not necessary. Its reflecting surface has an addition advantage in keeping the grain cool by radiating the heat.

4

Process of Farming

Farm System

Apart from the manual labour, the greatest amount of farm power in India comes from animals. Among the animals, bullocks are used most. The efficiency of using bullocks depends upon their feeding, maintenance, manner of yoking and training. The draught exerted by a work animal is related to its body-weight, the mechanics of working, the method of hitching the implement and the type of harness or yoke used. The following table shows the range of the ratio between the weights of the bullocks and the draught exerted in ploughing.

Body Weights of Bollocks and the Draught Exerted in Ploughing:

Average body-weight per pair of bullocks (kg)	*Locality*	*Draught (kg)*	*Weight: draught ratio*
432	Bankura	66.678	6.46
463.592	Tripura	46.449	10.0
453.592	Patna	53.9795	8.33
726.747	Jhansi	59.874	11.95

Contd...

Average body-weight per pair of bullocks (kg)	*Locality*	*Draught (kg)*	*Weight draught ratio*
771.107	Delhi	66.178	11.60
771.107	Aligarh	78.471	9.82
771.107	Coimbatore	146.149	5.20
907.194	Baroda	138.616	6.55
952.553	Karnal	134.535	7.07
1166.670	Indore	183.064	6.00

It may be assumed that the bullocks exert roughly 1/12 to 1/5 of their body-weights as draught. There is reason to believe that the draught animals generally have the ability to exert at least for a short period much greater pull than that indicated in the above table. There are cases on record, where a pair of bullocks weighing 725 to 900 kg has worked mouldboard ploughs having a draught ranging from 145 to 181 kg continually for 30 to 40 days without any adverse effect on their health. That is why sometimes a pair of good bullocks can do a better job than a 4- or 5-hp small tractor, which requires a very considerable amount of its horse power for its own movement. It cannot exert even for a small period appreciably increased draught which a pair of bullocks can exert.

Bullocks, buffaloes, or camels are commonly used in India as work animals. Camels are mostly used in desert areas and also for Persian wheels, oil *ghanis* or *kohlus* and to transport carts in Punjab, Rajasthan and U.P. Since bullocks are far commoner than other animals, some work has been done in determining their horse power. Their power is related to their body-weights. The following table shows the average body-weights of mature bullocks of the main Indian breeds used in farm operations:

Breed	*Average body-weight (kg)*
Nimari	303.907
Nagori	317.515
Siri	317.515
Sahiwal	408.233
Gaolao	408.233
Tharparkar	430.913
Kangayam,	522.601
Deoni	522.601
Haryana	508.023
Krishna Valley	544.310
Kankrei	544.310

Some tests have indicated that the horse power of two bullocks operating a Persian wheel is 0.877 and of those engaged in ploughing varies from 0.29 to 1.87. The walking speed of bullocks working in a circle with a radius of 3 to 4 m, while operating a Persian wheel is 2.5 miles per hour, with a draught of 50 to 51 kg. Bullocks engaged in ploughing walk at a speed of about 1.7 miles per hour and slow down to 1.2 miles per hour, at the end of the day's work. Thus at this speed, a pair of bullocks will plough 0.8 to 0.12 ha in a day with a *desi* plough and 0.2 to 0.4 ha in a day with a medium type of mouldboard plough in an alluvial soil.

Improvement in Yokes and Harnesses: It has long been realised that with improvement in the harness, the yoke and the method of hitching, the implements are likely to lead to a more efficient utilization of draught. Experiments in this connection are going on at Coimbatore, Allahabad, and Bardoli and a number of other places. The experiments conducted at Allahabad have shown that among the indigenous yokes, the Nagpuri yoke gives the greatest draught. This yoke has now been introduced, at least on an experimental basis, almost all over India and is being manufactured by some of the firms in U.P. The yoke is made of *babool* or *kikar* wood. Because of its shape and curvature, it rests properly on the necks of the bullocks and thus gives a greater draught with comfort to the animal. In case a good yoke is not used, the animal gets a neck gall and thereby reduces its efficiency. The weight of the yokes used in India varies from 7.25 to 15.875 kg depending on the type of wood used and the dimensions compatible with the strength required to withstand the pull exerted by the bullocks. However, most of the two-bullock yokes are good for a pull of 45.359 to 54.431 kg which is the draught commonly required for agricultural implements in use, particularly of the indigenous types. Efforts have also been made to introduce one-bullock yokes, but not with success.

Conservation Soil and Water of Irrigation and Drainage: Soil and water conservation go hand in hand. It is necessary to make the best use of the available rainfall in the scarcity areas, particularly in the districts, like Ahmadnagar, Rayalaseema and Sholapur. Some of the implements which are used for moving the soil or for making bunds have already been described. These implements include the soil scoop, the bund-formers, *keni,* etc. Their description, therefore, is not repeated here. The dry-farming implements, such as a long-bladed *bakhar,* a seed-drill having a distance of 3.81 cm between the tines and specially designed intercultural implements are recommended for dry-farming operations, the principal idea being to utilize, as far as possible, the largest amount of rain water. By bunding the fields and also the *nallahs,* some quantity of water is absorbed and conserved by the soil. This water, in turn, accumulates in open wells or tanks and can then be utilized for irrigation.

Another set of implements and machinery used for preparing the land for irrigation consists of a float, a drag, and a leveller. The float consists of a rectangular wooden frame to which cross plates are fixed. The last cross plate is inclined in the direction of the movement of the float. The float is mostly used where the land has already been levelled fairly well and where it is necessary to level it further in order that the irrigation water flows over the land uniformly and does not accumulate in depressions. A drag is an implement which throws the soil in one direction, making the channels flat-bottomed. The leveller is similar to the *keni,* but is longer for the uniform levelling of seedbeds. Water is carried through irrigation channels from wells to the fields and in order to minimize the loss of water through seepage, it is necessary to line them either with bricks or with cement. Recently, the workers of the Indian Road Research Institute have made a cheaper type of chemical out of creosote. This material can be used to plaster earthen water-channels. A kind of weedicide is also mixed with it to prevent weeds from growing in the channels.

One of the major difficulties of the farmers is the raising of water from the wells, tanks or canals to irrigate the fields. Since time immemorial, various indigenous devices have been used for lifting water, the simplest ones being a *dhenkli,* a Persian wheel, a *mhot,* a *pikota,* and *dons.* Where the water level is just near the ground, swing baskets are used. They can lift water from depth of 0.9 to 1.2 m. Two persons are required to work this swing basket. For lifts up to 1.828 m boat-shaped wooden or metal scoops, called *dons,* are used in Orissa, Bihar and Uttar Pradesh. The *dhenkli* consists of a lever and a bucket suitable for lifting the water from wells Up to 4.572 m from the surface of the ground. In the case of a *pikota,* which is common in Tamil Nadu, the operator has to walk on a levered beam. The capacity of the bucket of the *dhenkli* varies from 9 to 18 litres and its water duty, as investigated in Uttar Pradesh is less than a hectare.

The leather *mhot or charsa* is common in Bihar, Uttar Pradesh, Rajasthan, Madhya Pradesh and Maharashtra. The *mhots* in northern India are merely big leather bags, whereas the *mhots* in Andhra Pradesh, Maharashtra and Karnataka have a spout which allows it to be emptied automatically. They are worked by one pair of bullocks and up to depths of 9 to 12 m, but when the lift is more than 15 m, as in Rajasthan, two pairs of bullocks are used. The capacity of the bucket varies from 136 to 181 litres depending on the depth of the well. The discharge is about 6,819 to 9,092 litres per hour and it takes three to four days to irrigate 0.404 hectare. Some manufacturers have also prepared sheet metal *mhots* differing in design and capacity and incorporating into them small improvements like automatic valves in the bottom, self-balancing paired buckets or the bullocks operating in a circle instead of on a slope. The discharge per hour of the improved *mhots* is almost the same as that of the leather *mhots* used by the farmers.

The Persian wheels, or *rakats*, are extremely common in northern India, Uttar Pradesh, Rajasthan, Punjab and northern Maharashtra. Wherever the water level is between 4.5 and 9 m below the ground level, they are popular. A Persian-wheel consists of an endless chain and mild steel flats to which are fixed galvanized-iron buckets. The endless chain rotates on a large cage wheel. The axle of the cage wheel is geared with a pinion to the big gear, which is rotated by a pair of bullocks or a camel walking in a circular track. The modern Persian-wheels are fitted with ball-bearings, and several improved types have been produced. Mention may be made of 'Mayadas' double Persian-wheel which has found some favour in Uttar Pradesh. The average draft of a Persian wheel working up to a depth of about 4 m comes to 59 kg and that working up to a depth of 7.3 m, 86.1 kg. The discharge ranges from 8184.3 to 22730.4 litres per hour and the efficiency ranges from 34 to 44 per cent.

With the advent of engine-power or electric-power, a very large number of Indian farmers are going in for centrifugal pumps. This tendency is invariably growing where there are cash crops, or orchards or gardens. The centrifugal pumps are worked with internal-combustion oil-engines or electric motors. The state departments of agriculture also give *taccavi* loans, and other facilities for the installation of centrifugal pumps. In operating centrifugal pumps, priming is very important, because the centrifugal pump works on the principle of vacuum. The seals of the centrifugal pump should also be tight. In the case of those wells where the water level fluctuates considerably between the rainy season and the summer season and where it is necessary to apply irrigation in summer, additional platforms have to be installed inside the well for fixing the pump at different levels. Because a centrifugal pump has the limitation of raising the water up to the maximum level of 7 m theoretically and up to 5 m practically from the level of the water, the centrifugal pump has to be fixed at a height of 5 m from the water level to get the best out of it.

With the assistance of some of the foreign firms, the Indian manufacturers are also putting into market submersible or floating pumps that rise or go down according.to the fluctuating depth of the well. In such cases, it is not necessary to put one or two different plates for fixing the pumps. For very deep wells, the turbine pumps are recommended. These are manufactured specially on order. For tube-wells also, the turbine pumps are useful. In the case of filter-point tube-wells in the coastal region of Andhra Pradesh and Tamil Nadu, the centrifugal pumps can be used, because there the water level is, on an average, 5 m below the ground level. The filter-point tubewells fitted with Such centrifugal pumps are very cheap and, therefore, their use is becoming extremely popular in these regions.

The use of a hydraulic ram in the hilly areas is also recommended. It does not require any electricity or mechanical power. The energy that is developed by the falling of the water to a depth 2.438 m is used to raise it to the field level. Such hydraulic rams can be conveniently used in the hilly areas, such as Himachal Pradesh, Manipur, Mizoram and the Western Ghats. Irrigation water, even if it is readily available, should be judiciously used ; otherwise, it not only damages the crops but also damages the soil to the extent of making it barren. The injudicious use of irrigation water raises the salts in the soil to the surface, making the growth of plants impossible. It is only by the addition of organic matter or of gypsum or both that it is possible to reclaim such lands. Therefore, the extravagant use of irrigation water should be avoided. In those soils where water often accumulates, it is necessary to provide for drainage.

System of Draining

Good drainage is complementary to irrigation. In those areas where there is plenty of irrigation water, as in Punjab, Haryana and western Uttar Pradesh, drainage has become almost a national problem and is agitating the minds of the farmers and of the government. It is, therefore, necessary to know the methods developed for draining away the excess water from the fields. Water accumulates in the soil either from rains or through seepage from canals or irrigation channels and reservoirs. The high subsoil water-table damages agricultural land in many ways, leading particularly to the accumulation of undesirable salts in the root-zone or on the surface of the soil, involving drainage problems. The root cause of the drainage must be thoroughly investigated and understood. The problem of persistent water-logging is often complex and expert advice might be sought to solve it. In areas of heavy rainfall, good surface drainage has to be arranged. The drainage ditches should be wide and of trapezoidal or parabolic shape, so that they offer a rapidly increasing cross-section to the flow of the stream as the water level in them rises. On relatively flat land, they may be made to run down the slope. The velocity of water in the drains should be maintained at non-erosive level by proper grading and shaping. The spacing of the field drains should be worked out taking into consideration the character of the soil, the slope, the intensity of the rainfall, etc. A field drain should be laid to join a utility channel of adequate dimensions. These utility channels should be sufficiently big to take the maximum quantity of water.

In those soils where water accumulates underground within a depth of 0.6 to 1.2 m, underground drainage is needed. In case this high water-table is due to percolation from an external source, intersection drains should be installed to catch and divert the water. There are several types of underground drainage systems, but most of them are costly. As the work is of a permanent nature, it involves considerable expense. A master plan for draining the land should be prepared in consultation with specialist engineer and after making a preliminary survey of the levels. The best kind of drainage pipes are hard, burnt, earthenware flange-less pipes. The diameter of the pipe is usually 10 cm. The pipes should be laid well below the lowest root level of the crops proposed to be grown on the land. The joints of the pipes are important and should be left uncemented, but protected with a strip of tarred paper to prevent the entry of silt or soil, as the drainage of water takes place largely through the joints and not through the porous body as is usually thought. The spacing of the drains depends on the level of the land and the permeability of the soil. The drains can be placed in a grid or herring-bone fashion, depending upon the level of the water-logged area. In the case of sugarcane-growing areas in Andhra Pradesh and Tamil Nadu, the farmers are using mole drains. They are made at a depth of 0.45 to 0.6 m and are quite effective. The tractor-mounted trenching-machines are very useful in digging the channels for putting the tile drains.

In order to drain the water from the fields, the Japanese farmers use vertical pumps having low lifts, and high volume. Samples of such pumps were recently obtained from Japan and suitable designs were prepared. Such pumps are now available in India. The capacity of such pumps is 136,380 to 181,540 litres per hour and they can be effectively used for draining the soil, if the utility channels are available.

The Tractors

As explained earlier, even though we have to lay greater emphasis on manually-operated and bullock-drawn implements and machinery for the present, the slow mechanization of agriculture with power-operated implements has begun to come of age. It is generally considered that one of the main reasons for high agricultural yield in the Western countries is a high degree of mechanization of various farm operations. Mechanisation is possible there because of the shortage of labour and the overall development of engineering industries. So far, these two conditions have not existed in India. The labour has been relatively cheap and the industries are as yet, more or less, underdeveloped. However, these conditions have been rapidly changing since Independence. Agricultural labour, particularly at sowing and harvesting of crops is getting scarce. Almost any kind of agricultural implement or machine can now be manufactured in this country and that tool of good quality and at a reasonable price. The farmers are becoming aware of efficiently running a farm as a business. They are calculating the inputs and the resultant production, and agriculture is no longer a way of subsistence living. The governments of the states and the Centre are also doing their best to bring home to the farmers the results of research. These changing conditions necessitate the going in for mechanization and a fairly rapid progress is being made. That mechanization can succeed has also been shown by the farmers having already adopted diesel-engine pumping-sets as well as electrically-driven water-lifts. In certain states, quite a large number of farmers are using tractors and tractor-drawn implements. It may not be possible to substitute tractors for bullock power completely, but for the following operations, mechanized power can be profitably and effectively used and needs to be encouraged:

(1) For carrying out heavy jobs, such as summer ploughing or for reclaiming vast stretches of land. This work cannot be done either with manually-operated or bullock-drawn implements, as it needs lot of power. Many state governments have their own tractor units, which the farmers take on hire for deep ploughing, for reclaiming land and for carrying out soil and water-conservation practices, such as the bunding, levelling and terracing of the land and for good water management

(2) for eradicating weeds like *kans, hariali, dub* etc., which are deep-rooted and obnoxious ;

(3) for clearing jungles and for opening new lands ;

(4) for the desilting of tanks ;

(5) for big government farms, tea or coffee estates, and for using them on collective basis ; and

(6) for custom-hiring, wherever possible.

All the above types of operations have been done mainly through government agencies in the past. The best example is the Central Tractor Organisation of the Government of India, which functioned from 1947 till some years ago and reclaimed lakhs of acres in central India. Such large scale operations require heavy crawler-type tractors and specialized equipment, which are neither normally required by individual farmers, nor are they within their means.

Necessity of Tractor: For attachment to the tractor, we must provide the farmers with different types of implements to carry out different agricultural operations. Because of the higher horse power used, the tractors can do many more jobs which cannot be done by using bullock power. The construction of a tractor is similar to that of a motor-car in some ways, with the difference that the usual fuel used is diesel, the wheels are bigger and the speed slower. The construction of a tractor is sturdy, so that it can work in rough fields and arrangements have to be made for attaching different implements to it as well as for running stationary machines by using the power takeoff. The other phases of construction are similar to those of a motor-car, such as the engine, the clutch plates, the transmission gears, the differential, etc. Thus, one can see that a tractor is a modification of a motor-car to perform agricultural operations including bund-making, etc. The tractors can be classified into the following types:

Crawler Tractor: Such tractors are suitable for relatively high-draft and low-speed work. They can also be used on uneven ground or on soft marshy land where the wheels would not work. These tractors are heavier but their weight is more evenly distributed on a bigger area ; hence they do not compact the ground as much as the wheel-type tractors do. Their horse power ranges from 60 to even 500. They are employed for reclaiming large pieces of making the barren land fit for raising crops, for desilting tanks, for preparing barriers or bunds, for levelling the land and for draining away excess water.

Four-wheeled Tractors: Under this category come most of the agricultural tractors with an average horse power of 30 to 35. These tractors are fitted with pneumatic tyres. They can be used for ploughing, harrowing, sowing, harvesting, transport and belt work. They can also be used for intercultivation by making use of an arrangement by which the width between the wheels can be changed. This arrangement enables the tractor to be manoeuvred between the rows of growing crops. By using trailers, the tractors can be used for transporting manure, seeds, fertilizer, etc. These tractors are also being used for belt work such as for cutting chaff, lifting water and for working a wheat-thresher.

Small Size Tractors and Powar-tillers: A small tractor can be defined as machine having four wheels or two axles. A power-tiller is a machine having two wheels or one axle. The term "power-tiller" originated in Japan and most of the machines there are fitted with a rotary implement behind such a single-axles, two-wheeled small tractor. The horse power of the small machine ranges from 3 to 15. The Indian farmers are probably showing the greatest interest in such small tractors and power-tillers for the simple reason that the price of a good pair of bullocks is increasing alarmingly and that the bullocks require a lot of expense for their maintenance. The Indian farmers are looking forward to using small tractors or power-tillers in the range of 5 to 10 hp, which would not cost more than its 2,000 to 3,000. The main jobs that such machines should be able to perform are ploughing, harrowing, puddling, transport, pumping water and spraying insecticides. If the machines within the range mentioned above can do these jobs and are available for the price mentioned, they would certainly prove to be very popular. Such small tractors and power-tillers would also prove useful because of the reduced sizes of the land holdings. In most of the states the ceiling on land has been fixed at 10.117 to 12.140 ha and it is not economical for the farmers to use bigger tractors. The conventional four-wheeled tractors can be used on such lands only on co-operative basis or on custom basis. The four-wheeled small tractors have the riding arrangement, whereas the power-tillers are walking-type tractors. In Japan, these power-tillers have become extremely popular and about eight to ten lakhs of them are in use on Japanese farms. They are also commonly used in Germany, Italy, France, etc. The power-tillers are provided with a variety of attachments, which can perform almost any agricultural operation. Already, three factories in India have started producing power-tillers with Japanese collaboration and one more factory in the co-operative sector is coming up.

Apart from the types of tractors mentioned above, there are half-track tractors as well as baby track-layers. For wetland paddy-cultivation, cage wheels have been designed. They are also called skeleton wheels. They can be broad or narrow, ranging from 0.5 to 1 m in width. They help the tractor to distribute its weight in wetland cultivation and do not allow the tractor to sink. The cage wheels are also useful in puddling.

Tractor Maintenance: The tractor is a wonderful piece of invention, but as in the case of similar other machines, its maintenance is extremely important. Without maintenance, the machine will not last long, nor will it give the expected outturn. It is also a complicated machine. Much research has gone into its development and great care is taken in manufacturing it. When it is delivered to the farmer, every care is taken to see that it is in as perfect a mechanical condition as modern science and technology can make it possible. There is one important difference between a tractor and a team of animals. Once the animals are properly fed and housed, they can take care of themselves for most of the time. But a tractor is a piece of dead metal and it is necessary for the farmer to keep it in good adjustment and repair through careful preventive maintenance. There are four factors which affect the maintenance procedures of machinery:

(1) Machinery must be designed to suit the jobs and conditions they are expected to meet in the locality;

(2) the owners and operators should have a thorough knowledge of the machines, their operation and maintenance techniques;

(3) a thorough study of the instruction books and manuals, which are usually carefully prepared and are supplied with every machine that has been purchased from a reliable manufacturer or his dealer, is necessary ; and

(4) a good spare-parts dealership and delivery service are essential.

It is not possible to cover the entire subject of machine maintenance in a general chapter of this type. All that can be attempted is the discussion of a few important principles.

The fact that the engines of a conventional tractor and that of an automobile are in many ways similar has led to many prevailing misconceptions. A car-owner, who is accustomed to send his vehicle for lubrication to the workshop once a month is inclined to under estimate the importance of the lubrication in maintenance of a tractor, as he is under the impression that a tractor being a slower-speed vehicle than his car, should not need maintenance more than his car does. It must be realized that although the travelling speed is low, the static and dynamic loads on the engine transmission and other parts of a tractor are much heavier than those in the case of a car. Also, a tractor has to work for as long as eight to ten hours daily for days and pull heavy loads. This work is equivalent to driving a fully loaded automobile at 80 km an hour for ten hours a day for days together. Hence, what would be the monthly maintenance of an automobile would become a daily chore for the tractor. It is the ignorance of this factor more than that of any other that is responsible for many of the troubles that tractor-owners have experienced in the past.

In order to get the best out of a machine, principles of preventive maintenance may be defined as the technique of maintaining a tractor at the peak of its operational efficiency throughout its working life, keeping the wear and tear to the minimum all the time, i.e. during both operation and storage. The factor responsible for mechanical deterioration is wear and tear of the various components of the machine. Wear and tear can be classified as follows:

(1) Wear due to mechanical abrasion ;

(2) corrosion due to contact and reaction with moisture, air and products of combustion of fuel, etc. and

(3) the deterioration of certain materials susceptible to ageing, fatigue, etc.

Each part of the machine may be subject to one or all of the above factors to varying degrees. Preventive maintenance aims at minimizing and counteracting the effect of the various factors responsible for the deterioration of the machine. Conditions of operation also influence the frequency and intensity of maintenance schedule. Maintenance can be considered under the following three heads:

(1) Lubrication maintenance.

(2) Operational adjustments to keep the machine in proper operational tune for peak efficiency.

(3) Maintenance during storage and off-periods.

Maintenance of Lubricant: The effect of oil in reducing friction is well known. The present-day tractor is a complicated machine and has to be properly lubricated with the right lubricant at proper intervals. Mainly, the parts requiring attention are the engine unit, the engine accessories, the clutch, the transmission, the final drive and the chasis. Every tractor is supplied with a lubrication chart, stating the number and the location of the points requiring lubrication, giving further the type of lubricant and the frequency at which attention is required. Detailed instructions are given regarding how each job is to be attended to. It is important to see that meticulous attention is given to lubrication. A vast majority of the cases of machine breakdowns are traceable to faulty lubrication. The oils must be kept clean and changed at regular intervals. Several notable advances in tractor design have been made since the Second World War and many of the present-day designes are more compact, lighter and considerably more efficient than the older ones. Along with the improvements in the mechanical design, the lubricants and fuel technology have also advanced in no small measure, and they have contributed to the improvements in the performance of machines noticeable nowadays. Two decades ago it was a common practice to use a mineral oil straight as engine lubricant. But now, almost all the brands marketed by leading oil companies are blended with special additives to counteract certain undesirable and injurious chemical and physical changes that take place in the oil under the conditions to which it is subjected while the machine is being used. Some substances are added to it to dissolve the carbonaceous deposits that form on the essential parts of the engine. These deposits are removed when the engine-oil is drained. This is why the improved oils turn faster than the uninhibited oils. This darkening of the oil is an indication that it contains an active detergent which is actually dissolving harmful deposits.

It is incorrect, therefore, to conclude (as is ordinarily done) that the oil has gone bad prematurely. In fact, the colour of an oil is no criterion of its quality. The special protective substance used in an inhibited oil gets used up while the machine is operating. After sometime, the oil also becomes highly contaminated. It is, therefore, very important to change it at regular intervals. The oil specifications found in some manuals, formed the basis of selection some years ago. But these are now practically valueless. The best guarantee to a consumer regarding the quality of an oil nowadays is to buy it from an authorized and reliable dealer, preferably in factory-sealed containers.

Reclaimed Oils and its Use: Some persons—even those with technical background—advocate the use of reclaimed oils. Several firms manufacture the requirement needed for the reclamation of oils. That equipment is no doubt highly efficient, but it is however, suitable only in the case of straight mineral oils, as such an equipment can remove only the solid contaminants that are held in suspension physically, but cannot remove dissolved substances. For this reason, such equipment is unsuitable for oils containing chemical additives. It is also not possible through simple chemical tests to discover what proportion of the additive substances are still left unused in the oil. Therefore, no system of oil reclamation devised so far can deal with the new types of oils. The best thing for an average user of oils to do is to discard the spent oils drained from his machine. Some dishonest dealers offer to buy back these oils at attractive prices. But since the same oils may be used for adulteration, it is dangerous to be attracted by such a temptation.

The usual grade of oil generally used in engines is SAE 30, which refers only to the viscosity or the body of the oil and gives no indication of the quality of or additives present in it. Some air-cooled engines may use an oil as thick or heavy as SAE 50.

Lubricant Transmission: The present-day tractor transmission (gear-box) and the final drive system operate at much higher loads than formerly. The working temperature of the lubricant also can be very high and under heavily loaded operating conditions, the working temperatures may rise to 60°C or more. To withstand such severe conditions, the oils for use in such transmissions also have to be specially blended. Just any thick-looking oil will not do. These oils are refined and blended with the same meticulous care as the engine-oils and some special extreme-pressure additives are also added to give the oil some special qualities. Therefore, it is best to use the recommended grades and brands purchased from reliable dealers. The usual grade of oil specified for tractor transmissions and final drives are SAE 90 and SAE 140. Here again, the SAE specification number is no guide to the quality of the oil or to the additives present in it, since it is purely based on the physical property of viscosity. Transmission oils can also be contaminated and can deteriorate while being used and have to be changed once or twice a year. While working the machine, the oil level in the various components should be maintained at the correct levels by regular inspection and topping up. Overfilling should be avoided, as it is likely to lead to the failure of shaft seals and to other complications due to overheating and excessive frictional resistance.

Important Grease Points: Tractors are provided with several grease nipples to lubricate different parts. Greases are semi-solid lubricants of various consistencies designated as 00,0,1, 2, etc., higher numbers denoting progressively less viscous or heavier constituents. The main constituents in a grease are a metallic soap and a carefully chosen mineral oil. The main types of greases are a sodium-base grease, a calcium-base grease and mixed-base greases. The first type of grease contains soda-base soap, similar to household soap and it is water-soluble, but has a high melting-point. It is suitable for lubricating anti-friction bearings and is known as the bearing grease. The second type contains a calcium-base soap and is insoluble, but has a lower melting point. The so-called cup greases or water-pump greases fall into this category. The mixed-base greases have properties intermediate between the two and are called general-purpose greases. Greases should be carefully selected and extra care should be taken in its storage, as it easily attracts dust.

It is best to buy this lubricant in a size of packing that entails the minimum storage period under unsealed conditions. It is a very unwise policy to save money by going in for cheap brands of lubricants. The cost of the lubricant is insignificant in comparison with the cost of fuel and other expenses incidental to running a tractor. The contamination of lubricants with dust and dirt is responsible for highly accelerated rates of wear and tear. A well-maintained tractor may run for over 10,000 hours before it becomes necessary to undertake a major reconditioning job. A badly neglected tractor may be completely worn out after a few hunderd hours.

Efficient and safe operation of tractor. By proper operation, it is possible to attain economy in fuel consumption to the extent of at least 30 per cent or more. Proper attention to the routine adjustment of fan belt, valves, fuel and electrical systems will pay for the trouble taken and the expense incurred many times by increased efficiency and economy.

The majority of the tractors purchased nowadays are fitted with pneumatic tyres. These are perhaps the most expensive parts of a tractor. The most important factor affecting the life of a tyre is the maintenance of the correct inflation pressure, as stated in the instruction manual. Tyres should also he inspected periodically and any minor cuts should be promptly repaired. Nothing ruins a tyre so quickly and completely as running it in a deflated condition or even in an under inflated conditions. Therefore, the tyre pressures should be checked and maintained every day. A modern tractor is a very versatile piece of equipment and is provided with at least a belt pulley and power-take-off drive for operating various machines and devices. These devices can be highly dangerous if operated without proper guards and shields. Operators should not wear loose clothing. The best dress is close-fitting overall of *khaki* or dungaree cloth and a pair of heavy ankle boots.

Tractor-drawn Tools

The tractor-drawn implements can be classified into two categories. One set of implements is for tillage and the other for stationary operations. For tillage operations, the implements are ploughs, harrows, seed-drills, and combined harvesters and threshers. Under the stationary equipment are pumps, stationary threshers, winnowers, graders, etc.

In addition, the tractor can be used for transporting fertilizers, manures or produce by attaching a trailer to it.

The Plough: The tractor ploughs are of two kinds, namely, the mouldboard and disc ploughs. They cut two to five furrows at a time, depending on the number of complete bottoms or bodies composing the ploughs. In selecting a plough with the correct number of bodies in it, the maximum draw-bar pull of the tractor in the highest ploughing gear and the draught of the largest furrow normally required to be cut should be taken into consideration.

The correct speed of the tractor is important for good ploughing. The best speed for ploughing with normal types of ploughs is from 2.4 km/hour, depending upon the soil and other field conditions.

The depth of ploughing and burying trash. The best inversion is secured with mouldboard ploughs when the depth is less than two-thirds of the width of the cut of each bottom. A working depth of less than one-fourth of the bottom width would result in a poor quality of work. A 30-cm plough can work well up to depths varying from 8 to 20 cm. A digger-type plough cuts a furrow almost equal in depth to its width, but cannot deal so successfully with heavy trash. The shape of the share, particularly its vertical section, affects penetration.

A disc plough cuts a furrow whose width varies with the operating depth. The optimum depth is that which gives a width of cut equal to one-third the diameter of the disc. Unlike a mouldboard plough, the penetration is influenced also by the weight of the implement. So there is always some provision made for adding weight to secure the desired penetration in stiff soil. The shape of the discs, particularly their concavity, affects inversion and pulverization.

A mouldboard plough is more efficient than a disc plough if much of a surface vegetation has to be buried. But a disc plough is not liable to heavy damage from stones, roots and stumps.

The draft of a mouldboard plough varies from 0.140 to 0.21 kg/cm[2] of the cross-sectional area of the furrow slice in light soils to as much as 1.406 kg/cm[1] of extremely heavy and sticky clay soils. In summer in *kans* infested lands, the unit draft is as high as 13.607 kg.

For a 40-hp tractor, a three-bottomed plough, having 30-cm wide mouldboard and weighing about 453.592 kg is usually suitable. It will plough up to 20 cm deep and cover about 2 hectares a day.

The disc coulter is suitable for all types of ploughs ; it cuts cleaner and has a lighter draft than the knife coulter. The cut-away disc and the wavy-edged coulter are specially useful for working in a surface growth of creepers and tangled green-manure crops.

Disc ploughs are of two types, viz. disc ploughs and harrow or poly-disc ploughs. The disc ploughs are essentially heavy implements, having a few large discs (61 cm in diameter and above). They turn deep, 25 to 30.48 cm wide furrows and are best suited for cross-ploughing and stubble cultivation, particularly in hard soils. A trained worker can cover 1.6 to 2 hectares in an eight-hour day with a three-furrow disc plough or 2.24 hectares if the ploughing is light and shallow. For a 40-hp tractor, a plough having four 66-cm-diameter discs and weighing 16 to 861.825 kg is suitable. It will plough 25 to 28 cm deep.

The harrow ploughs are lighter, and have a large number of small-diameter discs for the same width of cut.

The field output of ploughs and harrows. The usual rule of thumb for estimating the output of an agricultural field operation is to multiply the effective working width of the implement in metres by the speed of the implement in kilometre per hour. The product gives the output in hectares in ten hours. This formula allows a reasonable idle time for turns on head lands, etc. Thus a 7.6 30-cm plough being pulled at a speed of 0.67 m/cm an hour can plough about 2 hectares

(3×1.5) in ten hours. It is useful to remember that a speed of 6.7 m/sec is equal to a speed of 24 km/hr. Thus if with a watch you find that a tractor moves 4.87 km/hr it will cover 2 km/hr. The average pace of an adult is 76 to 83 cm. It is a good thing to know the length of one's average pace by pacing a known distance. With some practice one can measure distance by pacing with sufficient accuracy for all practical purposes. The number of 33-inch paced in 20 seconds divided by 10 gives the speed in miles per hour.

Plough Care: The daily checking of plough adjustments, the tightening of bolts and nuts on the frame and bottoms and the regular attention to lubrication and replacement of worn-out parts prolong the life of the ploughs and lessen the cost of ploughing. To ensure efficient and economic working, the hitching of the ploughs should be in the line of the draught as nearly as possible. Furthermore, the hitch should be adjusted laterally, so that the tractor tyre does not rub against the furrow-wall unduly. While sharpening the shares, their original shape should be maintained. Cast-iron shares can be re-sharpened with a power grinder. Steel shares are sharpend by hammering them after reheating. Badly worn-out, dull shares increase the draught by 10 to 20 per cent and should be replaced with new ones.

For good ploughing, the following points should always be kept in view:

1. Furrows should be straight and compact, of uniform depth and width, and with even-finished ends.
2. The furrow wall and bottom should be clean-cut and the top of the inverted furrow slices should be level to give an even seedbed.
3. The depth of ploughing should suit the soil and the crop.
4. All vegetation, manure, etc. should be properly covered.

Farmers and Harrows: In general, cultivators are used for heavy work and for breaking large clods and the harrows are used for lighter operations.

Typs of Cultivators: The former have fine pulverizing effect, whereas the latter are good for deep cultivation in light soils. The average output of a 2.4 m-wide cultivator with 11 tines is about 0.6 hectares per hour. Heavy cultivators fitted with wide shares, going 5 to 8 cm deep even in hard soils, can be used for stubble clearance. Harrows are equipped with discs or spring tines or spikes. Disc harrows are used mainly for tillage in heavy or hard soils, but can also be used for a variety of other operations, such as the covering of seed and the destruction of weeds.

The spring-tined harrows are lighter implements, weighing about 11.339 kg per tine. They can work up to a depth of 13 cm and cover about one-third of a hectare per hour, with furrows spaced 30 cm apart.

The spike-toothed harrows are mostly used for light work and can be operated with small-wheeled tractors. The medium and heavy types can also be used for preparing seedbed on uneven and rough land, for covering seed and harrowing cereals.

Among rotary cultivators, small types carried on two wheels and operated by 5- to 8-hp engines are likely to prove useful for small holdings and market gardens. They produce the combined effect of ploughing and cultivating.

Heavier rotary hoes work up to 20 cm deep and need about 6 hp per 0.304 m width on average soils. Some efficient types of small tillers suitable for rice cultivation, have appeared in foreign markets. These are being tried in this country at present.

The Roller: On light land, a set of rollers, weighing up to 610 kg and 3 to 3.6 m wide, is required to provide full load for a small tractor. The roller needed for heavy land should weigh 864 kg more. The Cambridge roller with fluted surface is very effective as a clod-crushing unit.

Seed-drills and Planters: The seed-drills can sow all kinds of small seeds, e.g. wheat, oats, barley, *jowar* and paddy, and can be adjusted for variations in the size of seed, depth of seeding, seed-rate and the distance between the rows. Drawn by wheel-type or light crawler tractors, they can sow up to about 12 ha in a day, depending on soil conditions.

Planters are employed to sow seeds in rows far enough to permit mechanical intercultivation. Depending on the crop they are designed to sow, they are called corn, cotton, peas, bean, potato, and vegetable-planters.

The trailers are of two types, three to five-tonne fixed trailers or three- to five-tonne self-tilting trailers. Depending upon the time the trailers are to be emptied, the type of trailer should be selected.

How to Maintain Machinery ?: The Indian farmers are said to be not mechanically minded. It is necessary that this quality should be imbibed in them now. Unless this is done, they may not be in a position to work the implements and machines efficiently. One of the best ways of doing this is to get rid of the idea that for every different job a different village artisan is to be sent even for carrying out minor repairs and adjustments. The old sociological systems existing in the villages and based on different castes or professions must now be changed. For major operations, no doubt, specialized persons will have to be called, but for minor repairs and adjustments, a farmer must learn to do himself. By using a set of hand-tools, he can do the jobs requiring wood-cutting, some metal work, or sheet metal work, or even work in respect of leather or planting. Thus he need not sent for either a carpenter, or a blacksmith, or a tinsmith, or a leather-worker for minor jobs.

The following sample tools are recommended for every farmer using agricultural machinery:

A wood-saw, a small drill, a hammer, three sizes of chisels, a wooden plank, some rough files, an auger, pinches, a steel file, a hacksaw, a pipe wrench, an adjustable wrench, a one-foot piece of steel girder to serve as an anvil.

He should also maintain some spare screws, nuts, bolts, washers, etc. By keeping and learning to use such a set of tools and maintaining commonly required spare parts, a farmer would be able to carry out simple repairs, such as cleaning and oiling of chaff-cutters, cane-crushers, Persian wheels, sharpening ploughs shares and cultivator shares, the tightening of nuts, bolts, etc. The knives of the chaff-cutters should be sharpened by grinding or with an ordinary file. In the case of plough shares if they are to be sharpened by heating, they should be taken to a blacksmith. The ploughshares should be of high carbon steel. A well-sharpened ploughshare reduces the plough draft by as much as 20 per cent.

After the implements are used, they should be cleaned and oiled so as to retain the polish of the steel and also to avoid their rusting. They should be always kept in a shed.

The modern farmers have also to learn how to use improved implements. In the case of indigenous implements, the construction is so simple that it is not necessary to undergo any training, but in the case of modern implements, the farmers have got to learn their use. This training may range from a single day to about two weeks training in an organized way, depending upon the type and complexity of the machine. The rule to be remembered is that with every machine, the manufacturers supply a leaflet or a booklet containing instructions for operating and maintaining it properly. It should be thoroughly studied by the farmer and should be carefully preserved for future use or reference. The instructions in the booklet or manual should be followed in every detail. In these instructions, usually the type of lubricants to be used, their number, and the duration after which they should be used are also given. Therefore they are very important. All the bearings and other working parts should be oiled well to prevent their rusting, to secure efficient working and to prolong their life. Excessive oiling collects dirt which damages the machines. Only clean oil and grease should be applied and the oil holes should be covered. Agricultural machines are used seasonally. Therefore the farmers have plenty of time to repair or oil them. They should be stored in a roofed structure before they are required to be used in the next season.

Production of Agricultural Tools and Machines: Since Independence, quite a large number of firms in India have started manufacturing implements. There are about 150 organized industries in the country and they mass-manufacture agricultural implements.

The figure of 150 is not large and the village blacksmiths and carpenters still play a very important role in manufacturing simple agricultural implements. They will continue to do so for quite a large number of years, considering the extremely large number of farmers in India. They manufacture indigenous implements, bullock-carts, and nowadays they have also entered the field of repairing improved implements. If proper training is given to these blacksmiths and carpenters, they would prove very useful and effective in maintaining improved implements of the farmers. In order that they may shoulder their new responsibilities efficiently, the Government of India have established about 40 agricultural workshop wings attached to the extension-training centres for training the sons of blacksmiths and carpenters in repairing and maintaining improved agricultural implements. The trained artisans are also given loans for purchasing improved tools which they can use in the villages.

In the organized industries, most of the manufacture is limited to light mouldboard ploughs, turn-rest ploughs, ridgers, cultivators, hoes, wheel-hoes, olpad threshers, harrow *patelas,* diesel engines and centrifugal pumps, winnowers, sprayers, dusters, and sugarcane-crushers, tractors, power-tillers, etc. The implements are manufactured by using cast-iron, mild steel, high-carbon steel, sheet metal and wood. Most of the wood used is that of *babool* or *kikar (Acacia nilotica)* and *shisham (Dalbergia sissoo).* High-carbon steel is used for manufacturing shares of ploughs, shovels, chaff-cutter blades, pruning-knives, etc. Cast-iron is largely used for manufacturing the frameworks of sugarcane-crushers, chaff-cutters, shelters, bodies of turn-rest ploughs and other heavy parts. The plant-protection equipment, such as sprayers, dusters are made of copper, brass or brass sheets.

Considering the demand for improved implements and machinery, some of the state governments have established their own factories in the public sector. These factories are located in Haryana, Tamil Nadu, Uttar Pradesh, Maharashtra and Rajasthan. Most of them are now under Agro-Industries Corporations of the state governments. Other state governments are taking steps to establish at least one such factory in each state. Some of them may also undertake the manufacturing of power-tillers in the future. The difficulties of the manufacturers, such as the lack of particular types of materials, e.g., sheets and steel, the provision of factory sites to them in the industrial estates, the provision of sufficient electric power at a cheaper rate, need to be removed so that the manufacturing of implements becomes easy and profitable.

The Government of India have also undertaken the standardizing of improved agricultural implements, so that their main parts may be interchanged and uniformity may be maintained. Standardization also facilitates the inspection and quality-marketing. The Indian Standards Institution has already published about 50 standards for improved implements and tools and others are under preparation. The Board for Agricultural Machinery and Implements recently established by the Government of India has recommended a number of steps, so that improved implements can be designed, tested, manufactured and demonstrated to the Indian farmers as early as possible. Action is being taken on these recommendations.

Machine and Tool Use Practautions: Every precaution should be taken in the use of machinery and implements, particularly of the power-operated types, and those having sharp working parts. In the case of chaff-cutters, the wheels should be locked when not in use to avoid' any accidents. In the case of plant-protection equipment, the equipment should be thoroughly cleaned and dried after use and the poisonous insecticides should be stored safely away from grains and other articles of food. In the case of tractors, double riding should not be allowed. No operator should ride in between the tractor and the implement. The tractor operators should also be trained to use implements, such as ploughs, and whenever any obstruction occurs, the automatic release should be worked. Otherwise, there is a danger, that the tractor will tip. While operating the tractors and machinery, it is advisable to wear shirts and half-pants and no loose clothes, such as a *dhoti*, nor any flowing robes should be used. The fast-moving gears, transmission units, chains and spikes should be covered with guards. In the case of Olpad threshers, the discs should also be covered with expanded wire-netting. The tractor fuel, such as petrol, kerosene or diesel, should be stored very carefully. Like a small set of a mechanic's tools recommended for every farmer, a small first-aid box should be maintained. The trailers and tractors should be registered with the transport authorities and should be maintained as per rules and regulations.

It is a fallacy that the Indian farmer is slow at adopting new machines or that he does not need them. There are certain difficulties in the way of mechanization of agriculture, but they are not insurmountable. High prices of the machines mainly stand in the way of their adoption. To get over this hurdle, various methods can be adopted. Some of them have already been adopted in India.

(1) Custom-hiring by the farmers after they have completed their own agricultural operations.

(2) Custom-hiring by the agro-service centres established by the unemployed technical personnel.

(3) Custom-hiring by the Agro-Industries Corporations or other semi-government bodies.

(4) Hire-purchase.

(5) Collective farming.

(6) Starting of "Machinery Pools" as in Australia, or "Group Farms" as in Uganda in East Africa.

(7) Subsidizing the machines for the use of small and marginal farmers.

(8) Making credit available on easy terms through Land Mortegage Banks, or through I.D.A. Loans.

(9) Controlling the price of agricultural machinery and implements.

Agricultural engineering is slowly becoming a complicated subject, as more and more modern and sophisticated machines are being used. This is particularly true in the case of large land-levelling projects. It is time to recognize the utility and importance of the consultancy service in agricultural engineering and to make use of such services and expert advice while planning large and important projects. For speeding up the mechanization of agriculture, there is an urgent need for teaching the basic principles of farm machinery and implements to the boys in the schools. Therefore, simple books in regional languages on this subject need to be published for use in the schools all over India. The Government of India have approved a scheme for establishing a National Institute of Agricultural Engineering at Bhopal in Madhya Pradesh. This Institute will help to spread mechanization and make it more fruitful. A 'time will come when the mechanization of agriculture will take a big leap forward and we shall be able to help not only our farmers but also farmers in the developing countries of Asia and Africa.

5

LIVE STOCK

ELEMENTS OF NATURE AND BREEDS

Iron

- Approximately 65% of the total body iron is present in the form of haemoglobin.
- Widely iron is utilized in the body for various functions such as
    - (i) For transport of oxygen to the issue.
    - (ii) For maintenance of oxidative enzyme system within the tissue cells.
    - (iii) It is also concerned in melanin formation.

Sources of Iron: Various sources of iron like—Green leafy materials, most leguminous plants and seed coats are well sources.

Deficiency Symptoms of Iron

(i) By the deficiency of iron decrease growth rate.

(ii) Skin colour may be redden

(iii) The deficiency of iron is effect the formation of haemoglobin compound.

Zinc

- Zinc has been found in every tissue in the animal body.
- Sources of Zinc: The chief sources is wheat meal, safflower seed oil meal, fish meal etc.
- Deficiency Symptoms of Zinc:
  - (i) Growth retarded
  - (ii) Disorders of the feathers and hair coat.
  - (iii) Skin disease increase with the deficiency of zinc.
  - (iv) Disorders of bones increase.
  - (v) Reduced efficiency of feed Utilisation.

Copper: Even copper is very essential for the body and various functions in the body like for haemoglobin formation.

Deficiency of Copper

(i) Bone defects in grazing cattle and sheep on copper deficient.

(ii) Severe diarrhoea has been observed in many parts of the world by the deficiency of copper.

(iii) In sheep, copper deficiency produces a lack of black wooled sheep and characteristic less of "Crimp" form the fibers of all wool.

Manganese

- Even manganese are essential element for various functions in body.
- the highest concentration of manganese is in bone followed by pituitary gland, pineal gland, lactating mammary gland, liver gastrointestinal tissue, kidney and pancreas.

Deficiency Symptoms of Manganese

(i) In chickens the most dramatic mangnese deficiency syndrome occurs which is termed as "perosis" and is characterized by gross enlargement and malformation of the tibiometatarsal joint.

(ii) By the deficiency of Mn, reduced growth, Slightly reduced mineralisation, defective structure of bone.

Significance of Vitamins

- A vitamin is a organic compound which is present in normal food in minute amounts.
- Vitamins is essential for development of normal tissue and for normal health, growth and maintenance.
- When vitamins absent in the diet or not properly utilized, causes a specific deficiency disease or syndrome.
- Vitamins can not be synthesized by the host and therefore must be obtained either from the diet or from the micro-organisms of the intestinal tract.
- The vitamins are generally divided into two major groups: fat soluble which is usually found associated with the lipids of natural foods include vitamins A, D, E and K and water Soluble

Vitamins: Fat Soluble

- The members of this group are soluble in fat Solvents such as chloroform diethyl ether, petroleum ether etc. They are found in nature closely associated with those tissues that tend to store fats and oils.

Vitamin A

- Vitamin A was identified by Stechback.
- Vitamin A is dietary essential for all animals. Vitamin a does not occur in plants as such but it occurs as it precursor carotene. Carotene is known as provitamin A as it is converted into vitamin A inside the body.
- Vitamin A also known as retinol or retinal or retinoic acid.

Deficiency symptoms of Vitamin: various symptoms occurred in cattle and pigs like skin conditions, xerophthalmia through the deficiency of vitamin A. Poultry also affected by deficiency of vitamin A such as retarded growth, high mortality.

- Vitamin A is also responsible for the normal development of bones as it controls the activity of the osteoblast and osteoclasts of the epithelial cartilage.

Vitamin D

- Even vitamin D fat soluble.
- Vitamin D is also known as antirachitic factor as it prevents rachitis.
- Rickets is caused due to lack of normal calcification so that vitamin D is necessary for normal calcification.
- The disease rachitis, now commonly called rickets.
- Vitamin D is also involved to facilitate the clearance of phosphate in the kidney.

Deficiency Symptoms

(i) In older animals vitamin D deficiency causes ostemalacia, where there is reabsorptionof bone already laid down.

(ii) In young animals vitamin D deficiency results in rickets and retarded growth.

(iii) In Poultry, a deficiency of vitamin D causes the bones and beak to become soft and rubbery; growth is usually retarded and the legs may become bowed.

Sources of Vitamin D: Vitamin D is animal kingdom it is abundant in fish liver on entire body oil. Fresh green alfalfa forage has zero vitamin D value where as sun cured hay contain on an average 200 I.U. per 100 gms.

Vitamin E

- Mattil and Conklin (1920) indicated that rats fed on a milk diet supplemented with yeast and iron were unable to bear young.

- In 1922 Bishop and Evans announced the existance of a factor X in certain foods for normal rat reproduction.
- In the year 1936, Evans and his colleagues isolated pure vitamin E from the unsaponifiable fraction of wheat-germ oil.
- The active substance was later termed by Evans as "Vitamin E" and now at least eight compounds with E activity are known to occur in a variety of plant and animal tissues. These are called "tocopherols".
- Vitamin E acts as a antioxidant in the cellular level. Thus for example it prevents the oxidation of unsaturated fatty acids, mostly present in all cell wall components.
- Vitamin E also participates in normal tissue respiration.
- Vitamin E is also involved in the synthesis of ascorbic acid and co-enzyme and in the metabolism of uncleic acid (DNA) probably by regulating the incorportaion of pyrimidines into the nucleic acid structure and sulphur amino acid.

The Deficiency Symptoms of Vitamin E: Most animals fail to reproduce and young cattle and lambs muscle degeneration.

Sources of Vitamin E: Viamin E is widely distributed in live stock feed. It is found abundantly in whole cereal grains mainly in the germ and therefore in the by-products containing the germ.

- Leafy materials, green forage, good quality hay are very good sources.
- Alfalfa is very good sources of vitamin E.

Vitamin K

- Vitamin K was identified in 1935 by Henirk Dam as a factor present in green leaves which prevented a hemorrhagic syndrome observed in chicks maintained on a cow fat diet.

- Vitamin K is an essential metabolite for humans and for farm animals.
- Vitamin K helps in the formation of prothrombin which is necessary for normal blood coagulation.
- The deficiency of this factor lowers the prothrombin content of the blood.
- Prothrombin formation takes place in liver which is assisted by vitamin K.
- The prolongation of clotting time of the blood is due to lack of vitamin K.

Sources of Vitamin K: All green leaf plant materials, dry or fresh are rich sources of the vitamin.

Vitamins Soluble in Water

- Ascorbic acid was first isolated by Szent Gyorgyi (1928) from orange juice, cabbage juice and adrenal cortex; he named the compound hexuronic acid in recognition of the six carbon atom in the molecule
- Vitamin C is chemically known as L-ascorbic acid.
- Vitamin C highly soluble in water and slightly soluble in alcohol.
- Vitamin C is stored in all tissues with the highest concentrations in the adrenal and pituitary glands and lower levels in the liver, spleen and brain.
- Vitamin C is essential for the collagen formation.
- Vitamin C aids for the conversion of folic acid to its active form tetrahydrofolic acid.
- Vitamin C is also involved in the hydroxylation of proline, lysine and aniline which are important for normal physiology of the animal.
- It involves in the metabolism of lipids as blood cholesterol level appear to fall with the administration of ascorbic acid and rise due to deficiency of vitamin.
- It aid for the conversion of tryptophan to serotonin.

Deficiency Symptoms: Deficiency symptoms for form animals is unknown for man, monkey and guinea pigs extensive study has been made.

- A deficiency of this vitamin has been found to produce a disease known as scurvy which is characterized by Swollen, bleeding and ulcerated gums, loosening of teeth, weak bones and fragility of capillaries resulting in haemorrhages throughout the body.

Sources: Citrus fruits, tomatoes, green leafy vegetables, potatoes are certain other fruits and vegetables,

Vitamin B_1 (Thiamine)

- In 1926, Jansen and Donath, successor of Eijkmann succeeded in crystalizing vitamin B_1 in pure form. The identification of the structure of the vitamin and its synthesis was accomplished in 1936 by Williams and co-workers in the same laboratory.

Deficiency Symptoms: The deficiency of thiamine in humans occurs first as a numbness of the legs, later with pain in the calf muscle, severe exhaustion, finally emaciation and paralysis.

- The patient has difficulty in breathing, there is an abnormal enlargement of the right side of the heart and a decreases in the rate of the heart beat.
- Symptoms in animals termed polyneuritis are particularly characterized by unthrifitiness are paralysis convulsions and in birds like pigeons, chicks etc.

Paralysis appears to be due to accumulation in the brain and muscles of lactic acid.

Sources: Thiamine has widespread distribution in food but some food are very richest sources of this vitamin. In cereal grains most of the thiamine is in the outer grains layers. Green leafy vegetables bean are good sources. Animal products rich in thiamine include egg yolk liver, kidney etc.

Vitamin B_2 (Riboflavin)

- This is the IInd member of the vitamin B complex. This is the sparingly soluble in water but differ from thiamine in being comparatively stable to heat.

Riboflavin was synthesized independently by Khun etal (1935) at Heidelberg.

Functions: Riboflavin in the form of flavin monoumcleotides (FMN) and flavin adinine dinucleotide (FAD) acts as the prosthetic group of several enzymes involved in biological oxidation reduction reactions.

Deficiency Symptoms: The most characteristic symptom of riboflavin deficiency in chicks is leg paralysis or curled toe paralysis and other common symptoms are diarrhoea, low egg production and poor hatchibility in laying birds.

- In swine the symptoms are stiffched limbs, thickned skin, skin eruptions, lens opacities and cataract and exudate form the back and side.
- Pellagra is a multiple deficiency disease, therefore, pellagrins develop symptoms of riboflavin deficiency.

Sources: Riboflavin is widely distributed throughout the plant and animal kingdoms. Milk, liver, kidney and heart are excellent sources, many vegetables are also very good sources.

Vitamin B_4 and B_5 Niacin (Nicotinic Acid)

- An important disease known as pellagra, which was responsible for thousands of lives annually was caused due to the deficiency of nicotinic acid.
- People suffering from pellagra develop digestive disturbances characterized by diarrhea, fiery red tongue, ulcers in the mouth, dermatitis loss of appetite and other symptoms.
- In pigs, loss of weight, diarrhea, vomiting, dermatitis and normocytic anaemia are common symptoms.

- Nicotinic acid is also a dietary essential for adult fouel,
- In chickens its deficiency is accompained by poor growth, mouth symptoms, somewhat similar to that of black tongue in dogs.

Sources: Cereals and its by-products except corn are good sources. Groundnut oil is an excellent source. Alfalfa and other leafy material are fair sources, milk, coffee, yeasts are also good sources.

Vitamin B_6 (Pyridoxine)

- The vitamin was isolated by Keristezy (1938) and the synthesis of the vitamin was accomplished by Harris and Folkers in U.S.

Deficiency Symptoms (disease): In chicks, a deficiency causes acute convulsion, flatter on pan, usually starts kicking and generally die.

- In adult birds with mild deficiency hatchability and egg production are reduced.
- In rat characteristic skin lesion appear in the peripheral parts of the body such as paws, nose, ears, tails etc.

Sources: Good sources of vitamin include yeast and certain seeds, such as wheat and maize liver and green leafy vegetables.

Biotin (Vitamin H)

- Biotin is a relatively simple monocarboxylic acid, biotin is very slightly soluble in water and alcohol.

Deficiency Disease: In chicks biotin deficiency results in dermatitis similar to that occurring in pantothenic acid deficiency.

- By the deficiency of biotin various disease occurred like muscle pains, slight anaemia, anorexia, insomnia etc.

Sources: Richest sources of biotin are royal jelly, liver, yeast, molasses peanuts and eggs most fresh vegetables are fairly good sources.

Vitamin B_{12} (Cyanocobalamin)

- Vitamin B_{12} has been isolated in several different biologically active forms.
- Vitamin B_{12} anti-anaemia factor may play an important role in preventing and curing pernicious anaemia.
- It is a growth factor for chickens, rats, turkey and pigs, children are also affected in case of rat. Studies have shown that under certain conditions the vitamin is required for reproduction lactation as well as for growth.
- It is believed to be synthesized in rumen by sheep and cattle.
- Liver is the richest sources (natural sources).
- Alfalfa, pasture grass etc. also very good source of vitamin B_{12}.

Characteristics of Cattle Breeds

- Cattle belong to:

Phylum	—	Chordata
Class	—	Mammalia
Order	—	Artiodactyla
Family	—	Brovidae

Division of Indian Breeds of Cattle

(a) Classification of cattle breeds on the basis of types.

(b) Classification of cattle breeds on the basis of utility.

Indian Cattle Breeds

Milch breeds	*Dual purpose breeds*	*Draught-purpose breeds*
1. Gir	1. Haryana	1. Hallikar
2. Sahiwal	2. Ongole	2. Amritmahal
3. Red Sindhi	3. Kankrej	3. Khillari

Contd...

Milch breeds	Dual purpose breeds	Draught-purpose breeds
4. Tharparkar	4. Deoni	4. Bargur
5. Nimari	5. Nagori	
6. Dangi	6. Malvi	
7. Rathi	7. Ponwar	
	8. Siri	
	9. Gaolao	
	10. kenkatha	
	11. kangayam	
	12. Krishan valley	

Breeds of Milch Nature

1. Gir

- The origin of gir breed is gir forest of South Kathiawar.

Characteristics: Well proportioned body, ears are markedly long, the head is moderatly long but massive in appearance with prominent bony forehead, colour is seldom entire varying from almost red to almost black, spot of different colours are one the chief characteristics of this breeds. Average female weight is 386 kg while males weigh 544 kg.

Production: Cows are good milkers. Average yield 1746 kg. Bullocks are heavy and powerful.

2. Sahiwal

- The origin of this breed is central and southern dry areas of the Punjab particularly in Montgomery district in Pakistan.

Characteristics: Deep body, loose skin, short legs, stumpy horns, broad head and lethargic horns are short and thick. Various sheds of red, pale red and dark brown splashed with white colour.

Production: This is the best dairy breeds in India. In village the average yield is 2150 kg in 300 days lactation.

3. Red Sindhi

- The home of this breed is round about Karanchi and Hyderabad.

Characteristics: Medium size and compact, animals having well proportioned body, extremely docile, deep dark colour, heavy hump develop and sheath. Male weight average 450 kg while females 295 kg.

Production: Yield as high as 5,433 kg per lactation. Average yield 1.474 kg. Bullocks suited for field works.

4. Tharparkar

- The home of the breed is the Tharparkar district of Hyderabad.

Distinguishing Characteristics: Medium size, deep built, short, straight and strong limbs. Strong and well proportioned frame, broad poll and forehead is slightly convex with medium sized horns. The tail is fine with black switch.

Production: Cows are good yielders, bullocks suited for carting and ploughing. Average yield in village condition is 1474 kg.

Breeds: Dual Purpose

1. Haryana

Origin: The Haryana breed originated in east Punjab and are now extensively found in Rohtak, Hisar, Gurgoan etc.

Features: Proportionate body, head is carried high. Horns are short, curving upward. Popular colour is white or light grey. Long and narrow face, small and sharp ears, skin is of fine texture and close to the body. Legs are moderately long and lean. Tail is short, thin and tapering towards the end with a black switch reaching just below the hocks.

Production: Bullocks are good working animals for fast ploughing and road transport. Cow are average milkers with 1400 kg/lactation.

2. *Ongole*

Origin: Ongole tract of Andhra Pradesh.

Characteristics: Large, heavy, muscular, forehead is broad with stumpy. Horns thick at the base and firm without cracks. Hump is well developed and erect, filled up on both sides without leaning. Popular colour is white.

Production: Govt. Live-stock Research Farm Guntur has maintained this breed. Bullocks are powerful and suitable for cart and road work but are not fast cow are good yielder.

3. *Kankerj*

Origin: This breed originated in north Gujarat in the Mumbai province of India. Chief Characteristics: This is the heaviest breed, broad chest, forehead dished in the centre. Hump is well developed, tough skin, powerful body.

- In females colour markings are lighter.
- Tail is of moderate length with black switch.
- The animal is energetic and vigorous, excitable and nervous with strangers.

Production: Cows are fairly good milkers. Average yield is about 1333 kg but individual approach 3500 kg.

4. *Deoni*

Origin: North-West and western portion of Hyderabad.

Chief Characteristics: Resembles gir breed, less pronounced forehead. Colour is black and white or red and white with irregular patches. Deep chest, well arched ribs.

Production: Cow are good milker. Average yield about 900 kg in 300 days. Bullocks are well suited for heavy work.

Breeds of Drought Nature

Amritmahal

Origin: Karnataka State.

Cheif Characteristics: Compact form with short straight back, well arched ribs, narrow face and prominent forehead with a furrow in the middle long sweeping horns typical of Karnataka a type cattle. Tail is of moderate length with a black switch grey colour body.

Production: This breed is the best draught breeds of India. Cow are not so good milkers.

Kanagayam

Origin : This name of breed derived from Kanagayam division of Dharampuram Taluk of Coimbatore district in Tamil Nadu.

Chief Characteristics: Strong horns with sharp tips. Body moderatly long, straight back, short and strong neck. The colour of the bull is of grey with dark gray to black marking, while cow is white with black markings. Just in front of the fetlocks on all four legs, moderates sized hump, strong limbs.

Yield: The cow are poor milkers yielding about 666 kg per lactation. The bulls are excellent type for hard work.

Malvi

- This breed are found in malwa district in M.P. but now also found in Rajasthan.

Special Qualities: Deep short and compact body. Ears are alert and short in size, short but thick horn tapering to a blunt point, short neck. Tail is moderately longwith black switch. The general colour of the breed is grey to iron grey.

Yield: Bullocks are good for road and field work, good adaptability, cows are poor milkers 5-6 kg/day. This breed is essentially a good draught breed.

Siri

- These animals are found in hill part of Darjeeling, Sikkim and Bhutan.

Chief Characters: Massive body, small head, sharp horns, small ears, well placed hump covered with a tuft of hair at the top. Strong legs, colour is black and white or red and white.

Yield: Bullocks are well suited for cart purposes cows are poor milkers.

Hallikar

- This breed is mainly found in the district of Hassan in Karnataka state.

Chief Characters: This is the medium size animal compact and muscular in appearance. The head is usually long. The face is long with small ears. The hump is moderately developed.

Production: Cows are poor milkers and bullock are excellent draught type suitable both for road and field work.

Emergence of New Breeds

Karan Swiss

- This breed has been evolved at the NDRI, Karnal in Haryana by the breeding the Sahiwal cows with the frozen semen of brown Swiss bull.
- Cow body weight is 400-500 kg.
- Bully body weight is 600-750 kg.
- Ears are small.
- Neck is of medium size.
- Hump is long. The colour is light grey to deep brown.

Production: The lactation yield is 3355 kg in 305 days. The highest daily yield recorded is 43 kg.

Sunandini

- This breed is originated in Kerala by crossing the local non-discript cattle with Jersy, brown Swiss.

- The overall milk production per lactation is more than 2500 litres.

Karan Fries

- This breed also originated at National Dairy Research Institute, Karnal.

Buffalo: Domestic

- The domestic or water buffalo *(Bubalus bubalis)* belongs to the genus Bubalus.
- Now a day the buffalo is commonly known as an "Asian Animal".
- The domesticated buffaloes may be classified into two main categories
  1. The swamp buffalo
  2. The river buffalo

The Swamp Buffalo: The swamp type buffalo has an inherent partiality for water and love wallowing in water or mud pool.

The River Buffalo: The river type buffalo show preference for clean running water.

- The total world buffaloes population of 138.37 million is less then 10% of the total world bovine population.
- The annual rate of increase in buffalo population over the last 10 years was 1.7% compared to 0.7% in cattle.
- Largest concentration of buffaloes is in the Asia and Pacific countries.

Wallowing: Its Meaning

- Wallowing means rolling or floundering in mud or in water. The river breeds prefer the clean water of streams and pools punching close together.
- But the swamp buffalo of south-east Asia like to wallow in slime, in a mudhole accomodating one animal.

- Buffaloes apparently wallow for two main reasons
  1. To regulate body temperature and
  2. To control parasite.

Indian Buffaloes Breeds

Murrah group	*Gujarat group*	*Uttar Pradesh group*	*Central India group*	*South group*
1. Murrah	1. Surti	1. Bhadawari	1. Nagpuri	1. Toda
2. Nili Ravi	2. Jaffarbadi	2. Tarai	2. Pandhe-puri	2. South Kanara
3. Kundi	3. Mehsana		3. Manda	
4. Godavari			4. Jerangi	
			5. Kalhandi	
			6. Sambalpur	

- Buffaloes are classified on the basis of regions

Description

Breed of Murrah Group

Murrah

- The origin of this breed is mainly in Delhi and Punjab.
- Popular colour is jet black with white markings on the tail, face and extremities.

Production: The average milking capacity in well established herds is about 1400 to 2000 kg and butter fat content of 7% of milk in a lactation of 9 to 10 months.

Nili Ravi

- This type of breeds are found in the valley of Montgomery and Ferozepur.
- This animal is medium sized, deep frame, horns are small with high coil and the neck is long, thin and fine.

Production: She-buffaloes are high milkers average lactation period of 250 days. Males are used for draught work.

Gujarat Group Breed

Surti

- This breed is found in the south-western part of Gujarat state of India.
- The skin colour is black or brown and the colour of the hair varies from rusty brown to silver grey. The tail is long, this and flexible usually with a white switch.

Production: The average lactation yield of well-breed animals is 1600 kg with the fat content of 7.5%.

Jaffarbadi

- These breeds are found mainly in the gir forest of Kathiawar.
- Body is longer. Females are somewhat loose.
- Animal of this breed are generally black in colour. The Jaffarbadi males on the average weigh 600 kg. While females weigh 460 kg.

Production: Animals are heavy milkers ranging from 15 to 18 kg/day.

Mehsana

- The name of Mehsana derived from its home tract of the town Mehsana located in the north of the Gujarat state.
- Mehsana breed of buffalo was developed by cross of Murrah and Surti breed.
- The breed is a medium sized animal with a low-set deep body.

Breed Uttar Pradesh Group

Bhadawari

- These breeds are found in Bhadawari estate of Agra district and adjoining areas of Gwalior and Etawah.

This is found scattered in the surrounding of Yamuna and Chambal rivers.

- Medium sized and wedge-shaped body. Legs are short but stout. Colour is copper, hair scanty. Eyes are prominent, active and bright.

Production: Average yield from 2000 to 2070 kg in lactation period of 305 days with a high fat percentage. Males are used for drought purpose.

Tarai

- The name of this breed derived from the Tarai area of U.P. The breed is native to hilly area.
- Its moderate body, horns are long and flat with coils. The colour of the skin varies from black to brown.

Production: The breed is poor milk producer, which may be as low as 2-3 kg daily. Males are used for draught purposes.

Breed of Central India Group

Nagpuri

- These animals are mainly found in Nagpur in Maharashtra.
- Colour is usually black; occasionally white markings are observed on the face, legs and switch. Neck is also longer with heavy brisket.

Production: The female are moderately good yielders, producing on an average 1000 litres milk/lactation of 300 days with butter content between 7.0-8.5%.

Pandhepuri

- This breed is broadly found in south Maharashtra, parts of A.P.
- Animals are of medium size with long narrow face and very long flat and usually twisted with horns.

Production: Males are hardy and well suited for drought purpose.

Manda

- This breed are found in the hills above, Parlakimedi and Mandasa on the borders of Orissa and A.P.
- General colour of this breed is brown or grey with yellowish tuft of hair on the knees and fetlocks. The breed is medium sized animal. Eyes are sharp, horns are broad.

Production: Milk yield is satisfactory and males are hardy and can work in the hot sun.

Sambalpur

- The origin of the breed is Sambalpur area of Orissa.
- Animals are large and powerful having long narrow barrel, body colour is generally black.

Production: Females breed produce milk regularly and satisfactory. Males are active and good prefer for draught purpose.

Kalhandi

- The home of this breed is eastern part of A.P. and adjoining area of Orissa.
- Body colour of the animal is grey and ash grey colour. Eyes are prominent and large with narrow red margin around the lids. Tail is of medium length with white switch.

Production: Animals are hardy and docile. Milk yield of females is satisfactory. Males are used to carry load in hilly areas as well as for ploughing in plains.

Jerangi

- The breed is widely distributed in Jerangi hills of Orissa.
- This is the one of the dwarf breeds of buffalo and its height does not exceed four feet. The body colour is black.

Production: Females are poor for milk production but males are strictly used for ploughing in water logged paddy field.

Breed of South India Group

1. Toda: This breed is found in the Nilgiri hills of Tamil Nadu state.
2. Buffaloes the cows which become still weaker during the hot dry month (April-June). It is known as silent heat.
3. High buffalo calf mortality is one of the serious problem among dairy farmers. Various disease are responsible for more calf matality during winter season like pneumonia and II highest cause of calf is gastroenteritis. This disease may be due to viral, bacterial or parasitic infection. So good care are essential for devoid of this problem.
4. Services are given before feeding, as bulls seem to perform more quickly if this procedure is followed. If the interval between two successive, services is long the bull is allowed to serve a second time.
5. Buffalo bull attains sexual maturity later than ox bull.
6. The buffaloes are less distinguish in foraging and therefore consume a larger quantity of coarse fodder that is not readily eaten by cattle. This might be one of the reason why buffaloes thrive better than do cattle on coarse fodder.
7. During hot summer months, buffaloes need special comfort for which they love to have wallowing.
8. Buffalo bulls are more strong than the cow bulls. Fights between buffalo bulls are dangerous end in fatality.
9. Buffaloes are more susceptible to some disease like-rinderpest, haemorrhagic septicaemia than the indigenous cattle managed under comparable conditions.
10. India possesses the largest number of buffaloes (84.21 million in 1992). India is followed by China, Pakistan, Thailand, Philippines, Nepal, Indonesia, Vietnam etc.
11. Buffaloes have a low cost of milk production per litre which is a net outcomes of physiological superiority of buffaloes over cows.

- There are large-sized animals having long barrel and strong build. Horns of these buffaloes are variable in shape. Face is short and wide. The animals are not always docile.

Production: Females are good milkers, yield between 4.5 to 8.5 litres daily with an average of 7.0% fat.

Significance of Milk

- Indian buffaloes gives more milk than the Indian cow. Buffaloes have been contributing over 56% of total milk production. It has more total solids and the milk is richer in calcium and phosphorus and lower in sodium and potassium.
- Buffalo milk and ghee is white while cow milk and ghee is golden yellow in colour due to conversion of carotinoids into vitamin A.
- Buffalo milk is less suitable than cows milk for making Ghee and rosogulla.
- Breeding season of buffaloes is November to February because during this period bulls have been found to be very active sexually and quality and quantity of semen is very high and she buffaloes show maximum ovarian activity.
- The estrus cycle is about 21 days which divided into 3 phases:
  - (i) Pro-estrus—This may continue for 2-3 days.
  - (ii) Meta-estrus—The period of meta-estrus is 3-4 days.
  - (iii) Dio-estrus —This is the Last phase of estrous cycle 10-18 days.

Suitable Information About Buffalo

1. Mostly she-buffaloes show sign of heat at night (mid night).
2. The oestrus symptoms is weaker in.

6

LIVESTOCK CARE

- The word "Animal" means only those domesticated animals which are reared mostly for economic or for recreation purpose in any particular region. Such as cattle, buffalo, sheep, goat, yak, camel, dog, horse etc.
- The word "Husbandry" comes from the management of domestic affair but now presently it is also used in management of farming such as crop husbandry and animal husbandry.
- The term attached with animal husbandry includes proper feeding, breeding, health care, housing and many other activities.
- Definition: *"Animal husbandry" may be defined as a science as well as an art of management including scientific feeding, breeding, housing, health care of common domestic animals aiming for maximum returns.*

Present and Future Status of Animal Husbandry in India

- India is an agricultural country. At present 70% of total Indians are dependent for their livelihood on agriculture.

- According to FAO statistics on live-stock (1988), there are 201.4 million cattle, 76 million buffaloes, 57 million sheep, 108.57 million goat, 10.50 million pig, 205.98 million chickens.
- India has 14% of world cattle and 50% of the world buffalo population.
- Thus India is now occupying first position regarding its milk production among the countries of the world.
- The present contribution of live-stock to the national economy is estimated to be Rs 15,000 crore mainly from milk and milk products (70%), meat and meat products (11.5%). Poultry (8.8%) and dung for fuel (7.8).
- The total human population by year 2000 at the current growth rate would be around 997 million. To feed this population at the minimum recommended level.

Gestation Periods

Species	*Days*	*Species*	*Days*	*Months*
Live-stock		***Wild Animal***		
Ass	365	Ape, Barbary	210	
Cattle:		Bear, black		7
Angus	281	Bison		9
Ayrshire	279	Camel	400	
Browsn Swiss	290	Coyote	60-64	
Charolais	289	Deer Virginia	197-220	
Guernsey	283	Elephant		20-22
Hereford	285	Elk, Wapiti		8½
Holstein	279	Giraffe		14-15
Jersey	279	Hare	88	
Red poll	285	Hippopotamus	225-250	
Shorthern	282	Kangaroo, red	32-34	
Indian	282	Leopard	92-95	

Contd...

Species	***Days***	***Species***	***Days***	***Months***
Goat	148-156	Lion	108	
Horse, heavy	333-345	Llama		11
Horse, light	330-337	Marmoset	140-150	
Pig	112-120	Monkey, macaque	150-180	
Sheep:		Moose	240-250	
Mutton breeds	144-147	Musk ox		9
Wool breeds	148-151	Opossum	12-13	
		Panther	90-93	
Pets:		Porcupine	112	
Cat	59-68	Pronghorn	230-240	
Dog	56-68	Raccoon	63	
Guinea pig	58-75	Reindeer		7-8
Hamster	15-18	Rhinoceros		
Mouse	19-31	African	530-550	
Rabbit	30-35	Seal		11
Rat	21-30	Shrew	20	
Fur Animals:		Skunk	62-65	
Chinchilila	105-115	Squireel grey	44	
Ferret	42	Tapir	390-400	
Fisher	338-358	Tiger	105-113	
Fox	49-55	Walrus		12
Marten, European	236-274	Whale, sperm		16
Pine Marten	220-265	Wolf	60-63	
Mink	40-75	Woodchuck	31-32	
Muskrat	28-30			
Nutria (coypu)	120-134			
Other	270-300			

* Adapted from Merck Veterinary Manual, 5th edition.

Common Name of the Sexes, Young, Groups and Birthing of Various Animals

Animal	*Male*	*Female*	*Young*	*Group*	*Name for giving birth*
Antelope	buck	doe	kid	herd	kidding
Bear	boar	sow	cub	sleuth	cubbing
Beaver	boar	sow	pup	colony	—
Bird	cock/ stage	hen	fledgling/ nestling	flock	hatch
Bison	bull	cow	calf	herd	calve
Boboaf	tom	lioness/ queen	kitten	litter	—
Cat	tom	pussy/ queen	kitten	clowder	queening
Cattle	bull	cow/ heifer	calf	herd/ drove	calve
Chicken	rooster/ cock	hen/pullet	chick	flock	hatch
Deer	buck/stag	doe	fawn	herd	—
Dog	dog	bitch	pup/ puppy	kennel	whelp
Donkey	jackass	jeneet/ jenneyass	calt	herd	foal
Duck	drake	duck	duckling	flock	hatch
Elephant	bull	cow	calf	herd	clave
Fox	reynard	vixen	cub/pup	earth/ skulk	pupping
Giraffe	bull	cow	calf	herd	clave
Goat	billy/ buck	nanny	kid	trip	kidding

Contd...

Animal	*Male*	*Female*	*Young*	*Group*	*Name for giving birth*
Goose	gander	goose	gosling	gaggle/ flock	hatch
Hog	boar	sow/gilt	piglet/ shoat	herd/ drove	farrow
Horse stud	stallion/	mare/dam colt (male)	foal herd filly (female)	stable/	foal
Kangaroo	buck/ boomer	doe/flyer	joey	troop/herd	—
Lion	lion/tom	lionress/ she lion	cub	pride/ flock	whelp
Ostrich	cock	hen	chick	flock	hatch
Owl	owl howlet	jenny/ owlet	howolet/	flock	—
Ox	steer	cow	stot drove	herd/	calve
Rabbit	buck	doe	kitten	colony	—
Rat	buck	doe	—	colony	—
Seal	bull	cow	pup	herd/ harem	rokery
Sheep	buck/ ram	ewe/dam	lamb/ lambkin	flock/ hurtle	lamb
Swine	see hog				lamb
Turkey	tom	hen	poult	flock	hatch
Walrus	bull	cow	cub	herd	—
Whale	bull	cow	calf/cub/ pup	herd/ pod	pupping
Wolf	he-wolf	she-wolf	cub/pup	pack	whelp
Zebra	stallion	mare.	colt	herd	foal

- Animals are also classified on the basis of their nature of rumination (regurgitation). Some of them are listed below:
- List of common domestic animals classified on the basis of whether they are:

Non-ruminants (having single stomach)	*Ruminants (all ruminates; number of stomach may very from usual of four)*
1. Elephant	1. Cattle
Asian	Humped
African	Humpless
2. Rhinoceros	2. Baffalo
Indian	Asian
White	African
Black	Indian
3. Hippopotamus	3. Sheep
4. Horse	4. Goat
Domestic	
Wild	
5. Zebra	5. Nilgai
Grevy	
Plains Mountain	
6. Ass	6. Giraffe
7. Pig	7. Deer
8. Man	8. Bison
9. Dog	9. Llama (3 chambered stomach)
10. Monkey	10. Camels (3 chambered stomach)
11. Rabbit	
12. Hamster	

Mature Body Weights and Ages of Selected Species

Common Name	*Species Name*	*Mature Body Weight (kg)*	*Mature Age*	*Longevity (year)*	*Maximum Recorded Lifespan (yr)*	*Birth Weight (kg)*
Dairy cattle	*Bos taurus /indicus*	464-653	5 yr	20-25	30	25-45
Horse	*Equus caballus*	450	3-5 yr	20-25	50	45-55
Goat	*Capra hircus*	26-102	2 yr	10-15	18	2.5-3.5
Swine	*Sus scrofa*	70-128	—	16-18	27	1.0-2.0
Sheep	*Ovis aries*	34-80	2 yr	10-18	20	2.0-4.0
Rabbit	*Oryctolagus cuniculus*	3.0-4.9	1 yr	5-8	>13	0.04-0.08
Chicken	*Gallus domesticus*	1.5-3.3	20-30 weeks	2-5	30	0.049-0.065
Turkey	*Meleagris gallopavo*	5.0-14.8	28-35 weeks	-	12.3	0.52-056

Important Glossary

- *Animal Science:* The science which deals with the feeding, breeding and management practices of the live-stock.
- *Agalactia:* The failure to secrete milk following parturition.
- *Artificial Insemination:* This is the process in which semen with the live sperms assembled from the male and deposite in the female genetical tract by mechanical means.
- *Balanced Ration:* The combination of feeds that will supply the daily nutrient requirements of an animal.
- *Boar:* An uncastrated adult male pig used for breeding purpose.

- *Barrel:* It is the part of the animal body between the fore and rear legs.
- *Bleating:* The noise made by female goat at oestrus.
- *Breed:* Animal having a common origin and characteristics that distinguish them from other groups within the same species.
- *Bull:* An uncastrated male ox used for breeding purpose.
- *Bloat:* A disorder of ruminants characterized by the accumulation of gas in the rumen.
- *Broiler:* A chicken of either sex (male or female), usually of 8-10 weeks age that is tender meted flexible breast bone, cartilage, suitable for flying.
- *Bull Calf:* A male calf under one year of age.
- *Calf:* A young one of cow class (ox) of either sex.
- *Culling:* The process of eliminating non-productive or undesirable animals.
- *Crude Protein:* In the proximate analysis of feedstuffs, a figure representing protein content arrived at by multiplying the total nitrogen in a feed by 6.25.
- *Colostrum:* The first milk produced by the female immediately after giving birth to young. It is highly nutrious and a rich source of antibodies.
- *Cock:* The mature male fowl.
- *Capon:* A castrated male chicken.
- *Calving:* An act of giving birth to young one by a female of cow class.
- *Castrations:* Castration is the process or removal of testicles.
- *Colt:* A young male horse usually over one year of age.
- *Concentrates:* They are feeds which contain less than 18% of crude fibre and having more than 60% TDN. (HAS (Pre) 2004)

- *Cockaral:* A male fowl below one year of age.
- *Docking:* The removal of tail. This is the common practice in lamb.
- *Dry Cow:* A cow that is not producing milk.
- *Dry Period:* The non-lactation days between lactations.
- *Dual-purpose Animal:* One which is kept for both milk and meat production.
- *Doe:* A female goat, which has produced young one atleast once.
- *Digestible Energy:* The represented by that portion of energy consumed which is not excreted in faeces.
- *Domestication:* The process of adapting of life in intimate association with and to advantages of man.
- *Estrus:* The period of heat or sexual excitement in the female.
- *Estrous Cycle:* The period from one estrus or heat period, to the next.
- *Exotic:* The term used to describe animals foreign to a region.
- *Excrement:* The waste material discharged from the body, especially from the alimentary canal.
- *Ewe:* An adult female sheep which has produced a kid atleast once.
- *Fleece:* The entire coat of wool as it comes from the sheep or while on the live animal.
- *Forage:* Roughage of high feeding value. Grasses and legumes fed at the proper stage of maturity and quality are forages.
- *Feed Efficiency:* The units of feed consumed per unit of weight increase or unit of production.
- *Farrowing:* An act of delivery in swine.
- *Flock:* A group of the goat, it is called flock.

- *Food:* The edible substances which satisfied the instinct of hunger is food. Food for animal is known as feed.
- *Fertility:* It is the ability of an animal to produce large number of living young ones.
- *Foaling:* An act of delivering a young female horse.
- *Friedman Test:* A test for pregnancy in which a small amount of urine of the tested animal is injected into the blood stream of a virgin rabbit. Pregnancy is indicated by certain changes in the ovaries of the rabbit.
- *Gilt:* A young female pig not farrowed atleast once. She becomes a sow after farrowing.
- *Gross Energy:* The amount of energy released through the complete combination of feed in gred.
- *Geld:* A castrated male horse of any age.
- *Gestation Period:* It is the period between the day of conception to the day of calving.
- *Ham:* The rear quarter of a pig or of a pork carcase.
- *Hatching:* To bring forth young from the eggs by natural or artificial incubation.
- *Heifer:* A female ox of over one year, which has not calved, after calving she becomes a cow.
- *Hen:* A mature female chicken.
- *Hog:* A castrated male pig.
- *Herd:* The group of animals of cow class.
- *Hinny:* A hybrid whose sire is a stallion dam is a female donkey. It is also sterile.
- *Incubation:* Hatching of eggs of means of natural or artificial heat.
- *Impregnate:* To make pregnant to fertilize.
- *Intrinsic Factor:* A chemical substance in normal stomach juice necessary for the absorption of vitamin B12.

- *Kid:* A young one of goat either sex exceeding one year.
- *Lactation Period:* The number of days a cow secretes milk following each parturition.
- *Lamb:* Young one of the sheep of either sex male lamb known as ram lamb, female lamb known as ewe lamb.
- *Live Stock:* Those come in the class of mammalian viz, cow, goat, sheep, buffalo etc. which are pet are considered as live stock.
- *Mare:* An adult female horse, which produced a young one atleast once.
- *Milk:* The physiological secretion of the mammary gland of mammals which provides nourishment for young one is known as milk.
- *Mule:* A hybrid whose size is a donkey and is a mare it is usually sterile.
- *Mohair:* The wool of angora goats, a soft white wool in great demand for clothing.
- *Molting:* The sheeding and replacing of feathers of poultry birds.
- *Marbling:* Fat deposite in the lean tissue in meat.
- *Mastectomy:* The removal of the mammary gland.
- *Nutrient:* A compound that is required in the diet of a given animal to permit normal functioning of the life process.
- *Non-descript:* The animal of inferior quality that can not be identified as belonging to a specific breed.
- *Nymphomania:* An abnormal reproductive condition in cattle in which the female is in more or less constant estrus.
- *Ornithology:* A study of bird which are not as poultry is known as Ornithology.
- *Ovulation:* During the stage of metaestrus phase of estrous cycle mature follicle of the ovary ruptures and ovum is released. This process is known as ovulation.

- *Pashmina:* The under coat of kashmere goats, a soft wool in great demand for clothing.
- *Pork:* The meat that comes from swine.
- *Puberty:* A period of life at which the reproductive organs first become functional. This is characterized by estrus and ovulation in the female and semen production in the male.
- *Poultry:* Domesticated species of birds reared for egg, meat or feathers includes chickens, ducks, turkey etc.
- *Production Ration:* It is the food requirement of an animal in addition to the maintenance ration supply the nutrients needed for some forms of production.
- *Pica:* An appetite for materials not usually considered to be food such as is observed in phosphorus deficient animals.
- *Pinning:* The collection of dung around the vent of very young lambs that has dried to the point of interfering with normal bowel movements.
- *Rumination:* The process of regurgitation and rechewing food: unique to ruminants.
- *Reproductive Cycle:* The sexual cycle of non-pregnant female characterized by the occurance of estrus at regular interval.
- *Ram:* An adult uncastrated male sheep used for breeding purpose.
- *Ration:* It is often used interchangeable with diet but it may also mean a daily supply of food or feed.
- *Roughages:* They are bulky feeds containing relatively large amount of less digestible material.
- *Springer:* The term commonly associated with female cattle showing signs of advanced pregnancy.
- *Steer:* A male bovine castrated before development of secondary sex characteristics.

- *Still Born:* Dead at birth.
- *Semen:* Semen is a suspension of spermatozoa in seminal fluid, it is opaque white to light cream coloured fluid.
- *Sow:* An adult female pig used for breeding purpose.
- *Stallion:* A mature uncastrated horse used for breeding object.
- *Sire Index:* A figure that is indicative of the transmitting ability of a sire, in dairy cattle this figure is expressed in the terms of quantity of milk or milk component.
- *Staple Length:* The length of the cut hair or wool of a goat or sheep.
- *Stud:* A unit or herd of selected animals kept for breeding purpose.
- *Stripings:* Toward the end of milking, the last bit of milk is harder to milk out, the removal of this last bit is called stripping.
- *Super Ovulation:* The production by the ovaries of more than the usual number of eggs at the time of estrus. Usually induced by the hormone treatment.
- *Tassel:* Cartilaginous outgrowth at the base of the throat in some goat and sheep.
- *Tonned Milk:* It refers to milk obtained by the addition of water and skim milk. It should contain minimum 3% fat.
- *Tupping:* An act of matting male and female sheep.
- *Total Digestible Nutrients (TDN):* A term used in the animal feeding that designates the sum of all the digestible organic nutrients. TDN is a way of expressing the energy content of a feed.
- *Vasctomy:* It is an act of cutting down the portion of vas deferens and ligating it reaving the blood and nerved supply to testicles infect.

- *Wedder:* A castrated male sheep.
- *Weaning:* Taking the nursing young away from the mother and depriving it of the opportunity to nurse. This term is also used when calves are removed from diets containing fluid milk.

Significants Nutrients and Deficiency Symptoms

Nutrient	Deficiency Symptoms
Vitamins	
Fat Soluble	
1. Vitamin A (Retinol)	Xerophthalmia
2. Vitamin D (Cholcalciferol)	Rickets, osteomalacia
3. Vitamin E (Tocopherol)	Muscular destrophy
4. Vitamin K (Farroquinone)	Delayed clotting
Water Soluble	
5. Vitamin B-Complex	
Vitamin B_1 (Thiamine)	Polyneutritis (Beri-beri)
6. Vitamin B_2 (Riboflavin)	Curled toe paralysis
7. Nicotinamide	Dermatitis entritis
8. Vitamin B_6 (Pyridoxine)	Anaemia convulsion
9. Pantothenic acid	Goose stepping
10. Folic acid	Anaemia
11. Choline	Slow growth fatty liver
12. Biotin	Dermatitis loss in weight
13. Vitamin B_{12} (Cyanocobalamine)	Westing sickness
14. Vitamin C (Ascorbic acid)	Scurvy

Minerals	
1. Calcium	Rickets, osteomalacia, milk fever
2. Phosphorus	Rickets, osteomalacia, pica

Contd....

Nutrient	*Deficiency Symptoms*
3. Potassium	
4. Sodium	Dehydration
5. Chlorine	
6. Sulphur	
7. Maganesium	Hyperirritability
8. Iron	Anaemia
9. Zinc	Parakeratosis
10. Copper	Sway back
11. Manganese	Perosis
12. Iodine	Goitre
13. Cobalt	Coast disease
14. Molybdenum	Molybdenosis
15. Fluorine	Florosis
16. Selenium	Dangala disease alkali disease

Table: Number of Animals in India

Breed	*Number of animals (in millions) s*		*Growth rate*	
	1987	*1992*	*1887 & 1992*	*Annual mixture*
Cow	199.69	204.58	2.42	0.5
Buffalo	75.97	84.21	10.00	1.9
Sheep	45.70	50.78	11.20	2.2
Goat	110.21	115.28	4.60	0.9
Hog	10.63	12.79	20.40	4.1
Total (animals)	445.28	470.86	5.60	1.1
Poultry	275.32	307.07	11.50	2.3

- Meat:

1. Cow — Beef
2. Buffalo — Buffen
3. Pig — Pork
4. Goat — Chevon

Table: Milk, Egg and Wool Production

Year	*Milk Million Tonnes*	*Production Per capita availability gram/day*	*Egg Production Million number*	*Per capita availability number/ye*	*Wool Production (million kg)*
1950-51	17.0	124	1832	—	27.5
1955-56	19.0	—	1908	—	27.5
1960-61	20.0	124	2881	—	28.7
1968-69	21.2	—	5300	—	29.8
1973-74	23.2	—	7755	—	30.1
1980-81	31.6	128	10060	—	32.0
1990-91	53.6	176	21101	—	41.2
1995-96	66.2	—	27187	29	41.4
1996-97	69.1	202	27496	30	43.5
1997-98	72.0	203	28567	30	44.5
1998-99	75.0	211	30140	31	45.5
1999-2000	78.1	214	31320	33	46.1
2000-01	81	217	—	—	—

Sheep — Mutton

Poultry — Chicken

- Daily Nutrients Requirements of Cattle

(a) Maintenance requirements

Body weight (kg)	*Dry matter (kg)*	*DCP (kg)*	*TDN (kg)*	*ME (M.Cal)*
200	3.5	0.150	1.7	6.0
250	4.0	0.170	2.0	7.2
300	4.5	0.200	2.4	8.4
350	5.0	0.230	2.7	9.4
400	5.5	0.250	3.0	10.8
450	6.0	0.280	3.4	12.4
500	6.5	0.300	3.7	13.2
550	7.0	0.350	4.0	14.4
600	7.5	0.350	4.2	15.5

(b) Maintenance plus pregnancy requirements during last two months of gestation

Body weight (kg)	*Dry matter (kg)*	*DCP (kg)*	*TDN (kg)*	*ME (M.Cal)*
250	4.9	0.27	3.0	10.8
300	5.6	0.29	3.4	12.3
350	6.4	0.32	3.6	13.1
400	7.2	0.35	4.0	14.1
450	7.9	0.40	4.4	15.9
500	8.6	0.43	4.8	17.3
550	9.3	0.46	5.2	18.8
600	10.0	0.50	5.6	20.2

(c) Production requirements per kg of milk

Fat %	*DCP (kg)*	*TDN (kg)*	*ME (M.Cal)*
3	0.040	0.270	0.97
4	0.045	0.315	1.13
5	0.051	0.370	1.28

Contd...

Fat %	DCP (kg)	TDN (kg)	ME (M.Cal)
6	0.057	0.410	1.36
7	0.063	0.460	1.54
8	0.069	0.510	1.80

(d) Requirements of mature breeding (Bulls)

Body Weight (kg)	Dry matter (kg)	DCP (kg)	TDN (kg)	ME (M.Cal)
400	7.5	0.250	4.2	15.4
500	8.3	0.300	4.6	16.6
600	9.6	0.345	5.4	19.5
700	10.9	0.390	6.1	22.1
800	12.0	0.430	6.7	24.2

(e) Requirements of working bullock

Body weight (kg)	For Normal Work			For Heavy Work		
	Dry matter	DCP	TDN	Dry matter	DCP	TDN
200	4.0	0.24	2.0	5.0	0.25	2.7
300	5.8	0.33	3.1	7.0	0.42	4.0
400	7.6	0.45	4.0	9.8	0.57	4.8
500	9.4	0.56	4.9	11.2	0.71	6.4
600	11.2	0.66	5.8	13.4	0.82	8.0

Chemical Compositon of Milk of Different Species

Breed	Water	Fat	Protein	Total solid	SNF	Lactose	Ash
Ass	89.03	2.53	2.01	10.97	8.44	6.07	0.41
Buffalo	82.76	7.38	3.60	17.24	9.86	5.48	0.78
Camel	87.61	5.38	2.98	12.39	7.01	3.26	0.71
Cow	86.61	4.14	3.58	13.19	9.25	4.96	0.71
Goat	87.00	4.25	3.52	13.00	7.75	4.25	0.86

Chemical Composition of Different Milk By-products Percentage of Edible Part

Product	*Moisture*	*Protein*	*Fat*	*Carbohydrate*	*Ash*
Butter	16.0	—	81.0	—	2.5
Ghee	0.5	—	99.5	—	—
Curd	89.1	3.1	4.0	3.0	0.8
Lassi	97.5	0.8	1.1	0.5	0.1
Skimmed milk	92.1	2.5	0.1	4.6	0.7
Skimmed milk powder	4.1	38.0	0.1	51.0	6.8
Whole milk powder	3.5	25.8	26.7	38.0	6.0
Condensed/ sweetened milk	25.0	8.2	10.0	55.0	1.8
Evaporated milk	73.7	7.0	7.9	9.9	1.5
Chhana	57.1	18.3	20.8	1.2	2.6
Cheese	40.3	24.1	25.1	6.3	4.2
khoa	30.6	14.6	31.2	20.5	3.1

Important Institutions and their Location

1. CARI — Central Avian Research Institute, Izatnagar, Uttar Pradesh
2. CIRB — Central Institute For Research on Buffaloes, Hisar, Haryana
3. CIRG — Central Institute For Research on Goats, Mathura, Uttar Pradesh
4. CSWRI — Central Sheep and Wool Research Institute, Avikanagar, Rajasthan
5. IVRI — Indian Veterinary Research Institute, Izatnagar, Uttar Pradesh
6. HSADL — High Security Animal Disease Laboratory, Bhopal, M.P.

7. NBAGR — National Bureau of Animal Genetic Resources Karnal, Haryana

8. NDRI — National Dairy Research Institute, Karnal, Haryana

9. NIAGR — National Institute of Animal Genetic Research, Karnal, Haryana

10. NEBCAP — National Embryo Bio-technology Centre for Animal Production, Karnal, Haryana

11. NIANP — National Institute of Animal Nutrition and Physiology, Bangalore, Karnatka

12. NRCC — National Research Centre on Camel Bikaner, Rajasthan

13. NRCE — National Research Centre on Equine, Hisar, Haryana

14. NRCMP — National Research Centre for Meat and Poultry, Izatnagar, Uttar Pradesh

15. NRCMMP — National Research Centre for Meat and Meat Products, Hyderabad, A.P.

16. NRCM — National Research Centre on Mithun, Kohima, Nagaland

17. NRCY — National Research Centre on Yak, Dirang, Arunachal Pradesh

18. PDC — Project Directorate on Cattle, Modipuram Meerut, Uttar Pradesh

19. PDP — Project Directorate on Roultry, Hyderabad, A.P.

20. CIFA — Central Institute of Fresh Water Aquaculture, Bhubaneshwar, Orissa

21. CMFRI — Central Marine Fisheries Research Institute, Cochin, Kerala

Classification of Animal Diseases

- ***Bacterial Diseases***

1. Anthrax
2. Black quarter
3. Brucellosis
4. White scour
5. Paratuberculosis
6. Mastitis
7. Haemorrhagic
8. Pneumonia
9. Tuberculosis septicaemia
10. Erysipelous
11. Braxy
12. Enteretoximia
13. Foot rot
14. Fowl Cholera
15. Typhoid
16. Botalism
17. Pullorum

- ***Viral Diseases***

1. Swine fever
2. Pox
3. Foot and mouth disease
4. Plague
5. Blue tongue
6. New castle
7. Leucosis
8. Marek's
9. Bronchitis

- ***Fungal Disease***

1. Bronchomycosis
2. Histoplasmosis
3. Coccidiomycosis
4. Cryptococcidiosis
5. Rhinosporidiosis
6. Mycotic abortion
7. Mycotic mastitis
8. Ring worm
9. Brooder Pneumonia
10. Aflatoxicosis

- ***Protozoal Diseases***

1. Anaplasmosis
2. Babasiosis
3. Red coccidiosis
4. Theileriasis
5. Try panosomiasis
6. Try comoniasis

- ***Miscellaneous Diseases***

1. Tympanitis
2. Ketosis
3. Milk fever
4. Yoke gall

Specific Characters of Healthy Animal

Order	*Animal breed*	*Body temperature °C*	*°F*	*Heart beat/m.*	*Respiration rate/m*
1.	Cow	38.6	101.5	50-60	20-25
2.	Buffalo	36.8-39.4	98.3-103	40-50	15-20
3.	Sheep	38.9	102.0	70-80	20-30
4.	Goat	39.1	102.5	70-80	30-30
5.	Pig	39.2	102.6	70-90	10-20

Gestation Period and Body Temperature of Animals/birds

Animal	*Group is known as*	*Meat*	*Gestation period/ incubation day*	*Body temp. °F*
Cow	Herd	Calf-veal	283-285	101
Buffalo	—	Buffen	300-310	—
Sheep (Ewe)	Flock	Mutton	150	103
Goat (Doe)	Flock	Chavon	148	101
Swine (Sow)	Heard	Pork	114	101
Hen	Flock	Chicken	21	107

Foods for Livestock

Composition and Division of Feed Stuffs

- Live-stock feeds are generally classified according to the amount of a specific nutrient. They are divided into two general classes.

Roughages

- Roughages are bulky feeds containing relatively large amount of less digestible material i.e., crude fibre more than 18% and low in T.D.N. on air dry bases.

(HAS (Pre) 2003)

- Roughages are sub-divided into two major groups succulent and dry, based upon the moisture content.

  (a) *Succulent Feeds:* Succulent feeds usually contain moisture from 60-90%, succulent feeds are again classified into various types such as:

- *Pasture:* Pasture is the most convenient and economic for maintaining larger live-stock.
- *Cultivated Fodder Crops:* Even these are classified into two groups: leguminous and non-leguminous.
- Among leguminous fodders, cowpea, cluster bean are the most common kharif leguminous crops. They contain from 2-3% D.C.P. and about 10% T.D.N. on fresh basis and yield about 100 quintal of forage/acre. berseem and lucerne are two other leguminous crop in India which is grown in Rabi season. Lucerne and berseem contain an average 2.5 to 3.0% D.C.P. and 12% T.D.N. on fresh basis.
- Among non-leguminous fodder, jowar, maize and sudan grass (Sorghum sudanens) are most common kharif fodder yield range from 100-200 quintal/acre. Among the rabi non-leguminous fodder crops, oats and barley are most important of these two oat is far excellent for milch cattle.
- Non-leguminous perennial fodder crops consists of Napier grass, Hybrid Napier grass, Guinea grass, Para grass (Bracharia inutica). (HAS (Pre) 2004)
- *Tree Leaves:* The utilization of tree leaves for feeding to livestock is not common, they are used for feeding sheep and goats and are sometimes fed to cattle during periods of fodder crisis. The tree leaves and shurbs are generally rich in calcium but poor in phosphorus investigations conducted at I.V.R.I have shown that leafy fodder from the following species of trees and shrubs are suitable for use as maintenance ration for live-stock -pipal, babul, kachnar (Bauhinia variegata), bel, jharberi (Zizyphus numlaria).

- *Root Crops:* Root crops like turnips, sweders mongolds, fodder beet, carrot are used extensively in U.K. and other European countries for feeding during winter when other succulent fodder are not available. An important root crop in Southern India is Tapioca, which is grown extensively in Kerala. The yield per acre varies from 5-6 tonnes.

(b) *Silage:* The most economical method of raising live-stock is to feed then on grasses and legumes directly from the fields but the limit supply of there feeds at a uniform rate throughout the year. They are therefore, conserved as hay or silage for use at the time of crisis.

Dry Roughages

Hay: A method of conserving green crops is called hay making. The purpose of hay making is to reduce the moisture content of the green crop upto 15-20% to inhibit the action of plant and microbial enzymes. According to type of forages which are dried, hays are categorised as leguminous or non-leguminous.

- Among the leguminous plants the most suitable is lucerne, properly prepared lucerne hay contains 14-15% of D.C.P. and 50% T.D.N.
- Berseem and cowpea are more difficult to be converted into hay.
- Non-legume hays made from grasses are not as good feeds as legume hays. These are contain less protein, mineral matters and vitamins than legume hays.

Straws: Straws are poorest in protein and have the largest percentage of crude fibre. They are comparatively poor in phosphorus, in available calcium and also in trace elements.

- Straw consists of the stem and leaves of plants after removal of the ripe seeds by threshing and are produced from most cereal crops and from some legumes.

- Straw either from rice or wheat shall continue to be as one of the staple feeds for cattle and buffaloes throughout the rice and wheat producing areas of Asia because of:
  (i) Scarcity in the availability of good quality fodder.
  (ii) Straws are potential sources of energy for ruminants having 70% carbohydrate on dry matter basis.

Meaning of Concentrates

- A concentrate is usually described as a feed or feed mixture which supplies primary nutrients (protein, carbohydrate and fat) at higher level but contains less than 18% crude fiber with low moisture.
- On the basis of the crude protein content of air dry concentrates these are classified as either energy rich concentrates when CP content is less than 18% or protein rich concentrates when the CP content more than 18%.

Energy Rich Concentrates: These are described under the following categories:

(a) *Grains and Seeds:* The CP (crude protein) content of grains and seeds varies from 8-12%, which is deficient in lysine and methionine. The crude fibre content of the harvested grain is highest in oats and rice which contain a husk or hull formed from the inner and outer places.

- All cereals are deficient in vitamin D and in calcium but moderately rich in phosphorus and vitamin E.

(b) *Mill by Products:* The various cereal by products commonly used as live-stock feed like:

(i) *Bran:* Outer coarse coat of grain separated during processing e.g., rice bran, wheat bran.

(ii) *Flour:* Finely, soft ground meal of the grains consists primarily of gluten and starch from endosperm such as maize flour, wheat flour, etc.

(iii) *Hulls:* Outer covering of grain, generally not utilized as a live-stock feed.

(iv) *Germ:* It is the embryo of any seed. Wheat germ meal must contain at least 25% crude protein and 7% crude fat.

(v) *Meal:* Feed ingredient of which the particle size is larger than flour e.g., corn and oat meal contain protein between 9-18%.

(vi) *Polishings:* By product of rice, consisting of a fine residue that accumulates during polishing of rice kernals after initial removal of hulls and bran. It contains about 10-15% protein, 12% fat and CF between 3-4%.

(vii) *Groats:* Grain from which the hulls have been removed e.g., oats, rice etc. Protein percentage varies from 8-16% while fiber is between 1-3%.

(viii) *Shorts:* A by product of flour milling spring wheat consisting of a mixture of small particles of bran and germs, the aleurone layer and coarse fibre.

(c) *Mol Asses:* There are various type but all are concentrated water solutions of sugars, hemicellulose and minerals obtained usually as by products of various manufacturing, operations of the juices or extracts of selection plant materials:

(i) *Cane Molasses:* This is the by-product of sugar industry from which a maximum of sugar has been extracted. About 25-50 Kg of molasses results from production of 100 kg refined sugar.

- Cane molasses contains 3% protein, 10% ash.

(ii) *Beet Molasses:* This product is obtained as a by-product of the manufacture of beet sugar. Protein values are higher than cane molasses.

(iii) *Citrus Molasses:* When oranges or grape fruits are processed for juice, there remains 45-60% of their weight in the form of peel, rag and seed as wastes. The liquid obtained from pressing these wastes contains between 10-15% soluble solids of which 50-70% is sugar.

(iv) *Wood Molasses:* By giving high pressure at high temperature in presence of dilute acid, wood is converted to molasses.

- In the manufacture of paper, fibre boards, pure cellulose from wood, there results an extract which contains soluble carbohydrates and minerals of the wood material which may also be processed into molasses for live-stock feeding.

(d) *Roots:* A root crop consists of the fleshy underground parts of a harvested plants, grown primarily for its sugar content and is normally not given to animal as such e.g., turnips, sugarbeet, carrots etc.

- Root and tuber crop have traditionally been used more extensively as live-stock feed.
- In India, the only tuber which is used in large scale as live-stock feed is cassava tuber.

Protein Rich Concentrates

- Protein is the one of critical nutrients, particularly for young, rapidly growing animals and high producing adults, ingradients that contain more than 18% of their total weight in crude protein are generally classified as protein feeds.
- Protein supplements may be further categorical according to source of origin as:
  - (i) Plant proteins
  - (ii) Animal proteins
  - (iii) Non-protein nitrogen
  - (iv) Single-cell proteins

Nutrients of Feeding Stuff and Animal Body

- A nutrient is a substance that promotes the growth, maintenance, function and reproduction of an organism. The principle nutrients of all feeding stuffs are water organic and mineral matters.

Water and its Importance

Water is one of the important nutrient for all organism.

- We knew that animal may live for more than 100 days without organic food but they die in 5-10 days when deprived of water.
- Water Sources: Water available to an animals tissues comes from 5 sources:
  (i) Drinking water
  (ii) Water present in all feeds
  (iii) Metabolic water produced by oxidation of carbohydrate, fat and protein nutrients.
  (iv) Water liberated from polymerization reactions such as condensation of amino acids to peptides.
  (v) Performed water associated with the tissues which are catabolized during a period of negative energy balance.

ROLE OF WATER

1. Water act as a solvent action.
2. Water acts as lubricant to prevent friction and drying.
3. Hydrolytic Reaction: Hydrolysis is an important chemical process involved in digestion and other metabolism.
4. Transport: Water acts as a vehicle for various physiological process like:
   (i) For the transport of various food stuffs from place to place.

(ii) For the manufacture of various secretions, such as digestive juices etc.

(iii) For the drainage and excretion of the end products of metabolism.

(iv) For absorption of food material from intestine.

(v) For reabsorption from kidney tubules.

5. Body temperature is regulated by water.

   - Water intake by adult cattle is 3-5 kg for every dry matter intake, for suckling calves it is much higher being 6-7 kg/kg dry matter.
   - The consumption of water is increased during late pregnancy.

Meaning of Carbohydrates

- Carbohydrates are literally means hydrate of carbon.
- When sugar are heated for a long time in a test tube a black residue (carbon) and droplets of water condensed on the sides of the tube will be obtained.
- The carbohydrates are the major constituents of most plants comprising from 60-90% of their dry mass. In contrast, animal tissue contains very small (less than 1%) amount but without which life will be at stake.

Carbohydrates: Biological Significance

1. The main function of carbohydrate in the form of glucose and glycogen is to furnish energy for the body. More than 50% of the energy value of the diet is provided by the carbohydrates.
2. Some carbohydrates have highly specified functions e.g., ribose in the nucleo protein.
3. When the intake of carbohydrate is more than what is required by the body, the excess is utilised for the transformation of fats for the carbon skeleton of proteins.

Division of Carbohydrates

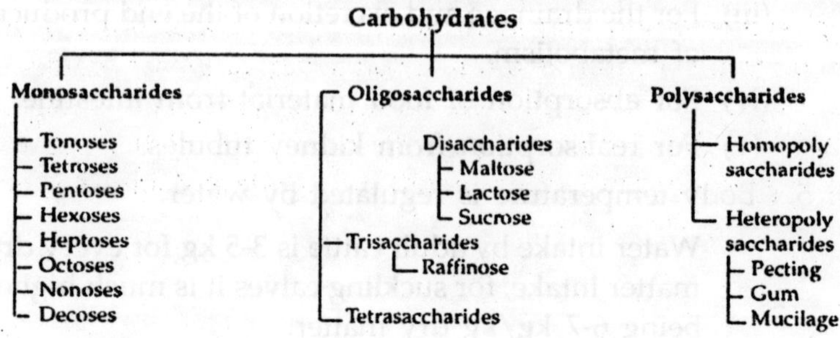

Meaning of Crude Fibre

- Crude fiber is also known as dietary fiber and includes a variety of polysaccharide substances viz, cellulose, hemicellulose, pectins and mucilages.
- Pectin and mucilages are not fibrous.
- A widely accepted definition is:

"The sum of lignin and the polysaccharides that are not digested by the endogenous secretions of the digestive tract".

- Crude fiber is available only in vegetative kingdom.
- Two types of crude fibre is: 1. Water soluble fibre, 2. Water insoluble fibre.

Nitrogen Free Extract

- A relatively soluble carbohydrates are classified as the nitrogen free extract (NFE) and include the mono and disaccharides plus the starches perhaps a part of the hemicelluloses, based on their relative solubility and digestibility and some cellulose and pentosans only with a limited amount of lignin.

  NFE percentage = 100 – Moisture% + crude protein% + ether extract% + crude fibre% + Ash%.

Meaning of Lipids

- The term lipid refers to any compound that is soluble in ether or benzene or in chloroform but only sparingly soluble in water.

- Fats are esters of glycerol that are solid at room temperature while oils are glycerol esters that are liquid at body temperature. 98% of animal lipid is fat. Waxes are esters of fatty acids with alcohols other than glycerol.
- Ration for adult ruminants should contain no more than 3-5% Fat and 15-20% fat for non-ruminants.

Division of Lipids

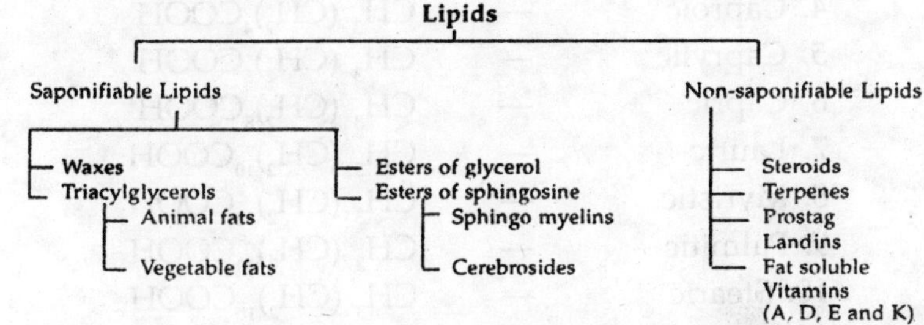

- The classification of lipids on the basis of chemical nature of individual lipids—
  1. Simple Lipids: Fats, oils and waxes are simple lipids- Fats and oils are esters of fatty acids with glycerol and waxes are esters of a fatty acids with alcohols other than glycerol.
  2. Compound Lipids: These are esters of fatty acids containing groups in addition to an alcohol and fatty acids. They include phospholipids are fats containing phosphoric acid.
  3. Derived Lipids: Derived lipids include substance derived from the previous groups as in serial 1 and 2 above by hydrolysis that is, fatty acids, glycerols and other alcohols.
  4. Miscellaneous Lipids : Such as sterols which are lipids with complex phenanthrene type ring structures.

Meaning of Fatty Acids

- Fatty acids are integral parts of lipids and may be divided into

(a) Saturated Fatty Acids

Saturated fatty acids have no double bonds in the chain.

- Their general formula is CH_3—$(CH_2)_n$—COOH
- Various saturated fatty acids given below:

1. Acetic — CH_3COOH
2. Propionic — CH_3CH_2COOH
3. Butyric — $CH_3\ (CH_2)_2COOH$
4. Caproic — $CH_3\ (CH_2)_4COOH$
5. Caprylic — $CH_3\ (CH_2)_6COOH$
6. Capric — $CH_3\ (CH_2)_8COOH$
7. Lauric — $CH_3\ (CH_2)_{10}COOH$
8. Myristic — $CH_3\ (CH_2)_{12}COOH$
9. Palmitic — $CH_3\ (CH_2)_{14}COOH$
10. Stearic — $CH_3\ (CH_2)_{16}COOH$
11. Arachidic — $CH_3\ (CH_2)_{18}COOH$

(b) Unsaturated Fatty Acids

- Unsaturated fatty acids have one or more double bond.
- Various unsaturated fatty acids given below:

1. Palmitoleic fatty acids — $CH_3(CH_2)_5CH{=}CH(CH_2)_7COOH$
2. Oleic acid — $CH_3(CH_2)_7CH{=}CH(CH_2)_7COOH$
3. Linoleic — $CH_3(CH_2)_3(CH_2CH{=}CH)_2(CH_2)_7COOH$
4. Linolenic — $CH_3(CH_2CH{=}CH)_3(CH_2)_7COOH$
5. Arachidonic — $CH_3(CH_2)_3(CH_2CH{=}CH)_4(CH_2)_3\ COOH$

- The term protein is taken from the Greek word proteus.
- The term protein suggested by Mulder in 1840, who clearly recognised that protein was necessary for life in a more fundamental way than could be attributed to either carbohydrate or lipid.

Role of Minerals

- Animal body contains about 3% minerals, which are constituents of animal tissues.

- In animal body there are about 30-40 mineral elements which occur largely in the various parts of their body.
- Minerals are very necessary for body growth milk production and another activities.
- Over 80% of the total mineral matter is found in the skeleton, giving strength and rigidity to the bones and teeth, the remaining mineral elements occur in the tissues and in the blood.
- Minerals are necessary in the animal body in general for the following reasons.

In Tissue Consisting and Repair

- For the formation of new bones and tissues in growing animals. Bones and teeth are high in mineral matter.

1. Composition of Fresh Bone:

Water	—	45%,	Ash	—	25%
Protein	—	20%,	Fat	—	10%
Calcium	—	36%,	Phosphorus	—	17%
Magnesium	—	0.8%,	Other's	—	46.2%

2. For the formation of hair, hoofs, horn.
3. A small amount is present in all soft tissue, which quantitatively may be small but are vital for life processing.
4. Blood cells also contain a small amount of minerals for the normal functioning of blood cells.

Minerals Act as Body Regulations

1. For the maintenance of proper osmotic pressure in the body fluids.
2. For the maintenance of neutrality of the blood and lymph.
3. To maintain a proper physiological balance between the various mineral ingredients in the blood and also for digestion.

In Milk Production

- To make good, loses of minerals secreted in the milk in milking animals.
- Cow's milk contains 5.8% ash or mineral matter on a dry basis.
- The description of various mineral elements given below:

Calcium: Calcium element is present in the larger amount in the body than any other cations. 99% of the body calcium occurs in the skeleton and teeth, the remaining 1% is vital for any normal animal body.

Calcium's Functions

(i) Bone formation including teeth and growth.

(ii) Clotting of blood *(Himachal Pradesh PCS (Pre) 2004)*

(iii) Maintenance of acid base equilibrium.

(iv) Regulation of heartbeat and working of muscle.

Calcium's Deficiency Symptoms

(i) Reduction in milk yield.

(ii) Thin shelled eggs with poor hatchability.

(iii) Milk fever in parturient can be due to sudden depletion of calcium.

(iv) Rickets in young and osteomalacia in adults leading to fragility of bones and fracture.

Sources: Legumes like alfalfa, sterilised bone meal and milk products are major source of calcium.

Phosphorus: Phosphorus is found in every cell of the body but most of the phosphorus is combined with calcium in the bones and teeth.

Phosphour's Functions

(i) Constituents of bone and teeth.

(ii) The oxidation of carbohydrate leading to the formation of ATP also requires phosphorus.

(iii) The constituent of the high energy compound ATP and thus is necessary for energy transductions essential for all cellular activity.

(iv) Phospholipids are constituents of all cellular membranes and are active determinants of cellular premeability.

Deficiency Symptoms of Phosphorus

(i) Phosphorus is required for bone formation a deficiency can cause rickets or osteomalacia.

(ii) Pica or depraved appetite has been noted in cattle when there is a deficiency of phosphorus.

(iii) In chronic phosphorus deficiency animals may have stiff joints and muscular weakness.

Sources: Grains, grain by products, concentrates like oil cakes, brans, milk product are the major source of phosphorus.

Potassium: Potassium is the staple cation of the intra-cellular fluid and plays a very important part along with sodium, chloride and bicarbonate ions.

Functions of Potassium

(i) Nerve transmission.

(ii) Maintenance of osmotic pressure.

(iii) Maintenance of acid-base equilibrium.

(iv) Heart beat relaxation

(v) It also aids in the uptake of certain amino acids by the cell.

Deficiency Symptoms of Potassium

(i) Weakness of the respiratory muscle

(ii) Weak extremities

(iii) Poor intestinal tone with poor intestinal distension.

Sodium: The sodium concentration within the cells is relatively low, the element being replaced largely by potassium and magnesium.

Sodium's Functions

(i) Maintains body fluid pH.

(ii) Regulates body fluid volume.

(iii) Takes active part in nerve functions and muscle contraction.

(iv) Functions in the permeability and carrier of the cells.

Deficiency Symptoms of Sodium

(i) Nervous disorder

(ii) Growth failure and reduces the utilisation of digested protein and energy.

(iii) Vascular disturbances.

(iv) In hens egg production is adversely affected as well as growth.

(v) Dehydration.

(vi) Corneal keratinization

Sources: All animal products, especially meat meals and foods of marine origin are richer sources.

Sulphur: Sulphur is present in all cells of the body. Primarily in the cell proteins containing amino acids, cystine and methionine etc.

The hormone insulin, the two vitamins biotin and thiamine also contain sulphur.

Sources: The main sources of sulphur for the body are the two sulphur containing amino acids such as cystine, cysteine.

Deficiency Symptoms of Sodium: The deficiency of the sulphur element is not usually considered in the body but the symptoms of sulphur deficiency is the same as deficiency of proteins.

Magnesium

Approximately 70% of the total magnesium is found in the skeleton. The remain amount is distributed in the soft tissues and fluids.

Magnesium's Functions

(i) This is the essential element for bone formation.

(ii) Another function of the magnesium is activate enzymes like phosphatases and the phosphorylation reaction involving ATP.

(iii) Controls the irritability of neuromuscular system.

Sources: Green fodder, bran, pericarp of cereal grains, cotton seed cake and lineseed cake are good sources of magnesium.

Deficiency Symptoms of Magnesium: Magnesium deficiency in adult ruminants a condition known as hypomagnesemic tetany. Associated with low blood level of magnesium.

7

CATTLE CARE

DAIRY FARM MANAGEMENT

- "The management is the art and science of combining ideas, facilities, processes, materials and labour to produce and market a worth while product or service successfully.

Functions of Management: Generally, a manager must perform 5 major functions-plan, organise, coordinate, direct and control.

Management Tools

- Some of the tools which are helpful in acquiring accurate information about our business and in guiding our management decisions are as follows:

1. Farm Records
   - (i) Complete inventories
   - (ii) Production records
   - (iii) Current expenditure and receipts
   - (iv) An annual production and financial summary
   - (v) Analysis of the year

2. Comparative budgeting
3. Annual budgeting
4. Partial budgeting

Caring for the Cattles

- Success in dairying depends largely on the proper care and efficient management of the herd.

Cow Caring

1. Usually a dairy cow will carry her calf a period of 282 days. However, they may range from 270 to 290 days after conception if accurate breeding records have been kept, which every farmer should do, the date can be calculated to within 1 to 10 days.
2. In handling advanced pregnant cows, care should be taken to prevent them from being injured by slipping on stable floors or by mounting cows. Separate the pregnant cows from rest and allow then to live in a little isolated way.
3. Symptoms that an animal is about to calve include swelling of the udder, swelling of the vulva and dropping away ligaments around the tall head. At this stage she should be housed in calving pen. The room should be clear, well ventilated, well bedded and finally, disinfected. Alternatively a small well grassed pasture free from trash or manure and close to the farmstead make a good calving place except during monsoon and cold month.
4. The majority of domesticated animals require little or no assistance in the actual act of parturition, provided they are in a reasonably healthy and vigorous state. At the first sign of calving, the front feet of the calf should appear first then the nose. Any abnormality in presentation requires immediate attention by a veterinarian. Remember that if the labour prolongs for more than 4 hours, abnormal presentation is probable, immediately provide veterinary aid.

5. Keep the cow warm to prevent her from chill and it is desirable to give her warm water.
6. After parturition the exterior of the genitalia the flanks and tail should be washed with warm clear water containing some crystals of potassium permagnate.
7. It is normal for the udder to become large and swollen just before calving. Controversy exists as to whether or not the udder should be milked out before calving. Special precautions should be exercised to see that old nails, loose glass pieces etc. do not cut and injure the swollen udder. Milk the cow partially to avoid milk fever after parturition.
8. The placenta will normally leave the cow within 2-4 hours. If it is not expelled between 8-12 hours, administer ergot mixture. Beyond 12 hours, apply manual help by a veterinarian.
9. There are always dangers that high producing cows will develop milk fever and mastitis. The dairyman should remain alert for any symptoms of the disease to avoid milk fever. It is best not to draw all the milk from the udder for a day or two after calving.
10. Feed the cow at first only bran mash moistened with lukewarm water to provide laxative effect. Some green grass may also be given after 2 days a mixture of oats, bran and lineseed mash can be used to replace the bran mash. If the cow is in good condition at the time of calving the amount of feed during these two days does not matter the amount of concentrates should then be gradually increased with the aim of reaching full dosages in two weeks.

Dairy Cattle House

Location of Dairy Building

- Points give below
  1. Topography and drainage
  2. Exposure to the sun and protection from wind

3. Accessibility
4. Water supply
5. Durability and attractiveness
6. Labour
7. Surroundings
8. Marketing
9. Electricity
10. Facility, labour and food

Kind of Housing

(a) The loose housing barn in combination with some type of milking barn:
   - This system is more economical. Some features of loose housing system are as follows:
     (i) Cost of construction if significantly lower than conventional type.
     (ii) It is possible to make further expansion without much changes.
     (iii) Facilitate easy detection of animals in heat.
     (iv) Animal feel free and therefore, proves more profitable with even minimum grazing.
     (v) Animals get optimum exercise which is extremely important for better health and production.
     (vi) Overall better management can be rendered.

(b) Traditional dairy barn
   - This method is costly and it is now becoming less popular day by day.
   - The following barns are generally needed for proper housing of different classes of dairy stock on the farm.
     (i) Cow house
     (ii) Calving box

(iii) Isolation box

(iv) Sheds for young stocks

(v) Bull or bullock sheds

Raising the Dairy Calf

1. Feeding and management of the calf before birth.
2. *Care of the calf after birth:* After birth the care of the calf is necessary. Here two things are essential. First is treatment of the calf and second is treatment of the navel cord.
3. Feeding calves

   (i) Feeding coloustrus

   (ii) Treatment of the calf to drink

   (iii) Feeding whole milk

   (iv) Feeding skim milk

   (v) Feeding calf starters

   (vi) Feeding grain mixture

   (vii) Feeding silage

   (viii) Providing Minerals: Milk is the very good sources of calcium and phosphorus.

   (ix) Supplying the antibiotics.
4. *Housing the Calf:* The object of housing is to provide shelter to the calves against sun, rain and other inclemencies of weather.
5. *Dehorning the Calf:* This is the process by which the horns of an animal are removed after birth by treating the tender horn roots with a chemical, mechanical or electrical dehorner.

Dehorning: Merits

(i) Dehorned animal will need less space in the sheds.

(ii) Horned animals are a danger to the operator.

(iii) Dehorned animals can be handled more easily.

(iv) Cattle with horns inflict bruises on each other that may result in heavy economic losses.

Demerits of Dehorming

(i) Animals with a nice horn have a style.

(ii) Animals with horn can defend themselves.

(iii) Some breeds have got an important indentification marks for horn.

Dehorning Methods

(a) Chemical Method: The chemicals commonly used are caustic soda.

(b) Mechanical Method of dehorning:

(i) Clippers and Saws: When older cattle are to be dehorned, especially designed clippers or saws are used.

(ii) Rubber Bands: Some farmers have reported successful dehorning of older cattle by using the rubber band method.

(c) Electrical method: Electrical method of dehorning are seldom used in India. The rod is heated by electricity and has an automatic control that maintains temperature at about 1000°F. Applying the electric dehorner to the horn button for 10 seconds is sufficient to destroy the horn cells.

6. Marking the calf.
7. Removing extra teats.
8. Castrating the Bull Calf: Castrating is the unsexing of the male or female and consists in the removal of both testicles or ovaries respectively. Its objects are to prevent reproduction, to increase faster gains, to produce a more desirable type of meat.

There are three methods of castrating a bull calf which are as follows:

(i) By making an operation in the scrotum where the vas deferens are disconnected from the scrotum. Thus the spermatozoa will not be able to flow out of the penis.

(ii) Castration with the help of a Burdizzo's castrator. The method is also known as "bloodless castration".

(iii) Recently a new method has been developed in western countries where they use a strong and tight rubber ring around the cord at the very early age of calf.

9. Preventing and controlling disease in calving.

Cattle's Habits

- Certain bad habits are prevalent among the cattle such as suckling another cow or herself kicking during milking, fence and rope breaking etc.

Some bad habits are given below:

Suckling

- Every dairyman has had experience with cows that such themselves or other cows, thus causing losses of milk, contamination of the udder and sometimes indigestion of the animal concerned. To prevent of this problem, the cow should be separated from the rest of the herd. A cradle or bull ring is put in the cow's nose and then two or three other rings are attached to it.

Licking

- Some animals, specially calves gets into the habit of licking other calves during the milk feeding period. This ultimately leads to the indigestion of hair which get entangled with the curdled milk in the stomach and forms of hair balls. This is serious problem. One of the precautionary method is to rub a pinch of salt or mineral mixture. On the tongue of the calf after each feeding. Repetition of this system will enable the calf to forget this habit.

Kicking

- Many heifers or cows kick when they are milked. It may be mainly because of handling by an unskilled milker. Before applying any remedial measure, it will be wise to search for the reasons of such habit. It may be possible that the cow is suffering from some serious disease of udder or other the control of this problem, in the severe cases anti-linking chains can be used. A clamp fits over each hock and a chain fastens them together.

Fence Breaking

- Some animals have the habit of breaking their fence of the enclosure in which they are grazed on jumping over the fences. The habit is formed due to the feeling that on the other side of the fence the grasses are more green or plenty. There is little that will stop a roguish cow except proper arrangements and good fences.

The Pigs

- Swine belong to the phylum Chordata, class Mammalia and family is Suidae.

Some Common Terms in Relation of Pig

Details		*Expression*
1. Species of pig is called	—	Swine
2. Group of animals	—	Drove or herd
3. Adult male	—	Boar
4. Adult female	—	Sow
5. Young male	—	Boarling
6. Young female	—	Gilt
7. New born one	—	Piglet
8. Castrated male	—	Hog
9. Sound produced	—	Grunting

Swine Breeds

- Recently there are about 60 recognised breeds of domestic pigs in the world. All have 38 somatic chromosomes.

Large White Yorkshire

- This breed is popular English Bacon breed which had its origin nearly a century ago in Yorkshire and neighbouring countries in northern England.
- Entirely white in colour, head is moderately long, face slightly dished, wide between the ears.
- Neck long, fine and proportionately full to shoulders with wide and deep chest.
- Back long, level and wide from neck to rump. Legs straight and well-set level with the outside of the body with flat bone, skin fine and white.
- Average sow weights from 250-350 kg.

Middle White Yorkshire

- This breed was evolved in Yorkshire of Northern England by the crossing the large white with smaller breed of Yorkshire extraction.
- This breed widely accepted as an excellent pork pig. This breed is hardy, grows rapidly.
- The colour of this breed is white. Head moderately short, straight, jaw neck medium long.
- Back long and level to root of the tail with well spring ribs.
- Legs straight and fairly short.
- Mature boars generally weight 270 to 360 kg.

Berkshire

- This breed originated in Berkshire (South-Central England). This breed is now highly valued as producer of good quality pork.
- The colour of this breed is black with 'six white points" that is white on feet, nose and tail.
- Short head with dished face and its ears are erect.

- Long back, straights legs, full grown boars weigh between 275-375 kg and sow weigh from 200-290 kg.

Landrace

- The origin of this breed is in Denmark.
- This breed is white in colour, although black sink spots.
- The carcass is more lean than that of the meat. There is less back fat and lord.

Tamworth

- The origin of this breed is central England. Golden red colour of this breed. Extremely good bacon type. Smooth sides and stong backs. The head is strikingly long and narrow with a long snout . The mature boars weigh upto 350 kg and the mature sows from 280 to 300 kg.
- Particularly this breed has been widely used for breeding purpose in the tropic countries.

Duroc

- This breed is moderately red coloured. The Duroc is noted for excellent rate of gain and feed efficiency. The carcass is considered as a good meat type. The weight of mature boar is about 400 kg and sow weight normally is 350 kg.

Chester White

- The breed had its origin in the south-eastern Pennsylvania.
- The breed is white in colour with some bluish spots sometimes found on the skin chester. White sow are very prolofic.

Hereford

- It is one of the new breeds of swine. The breed was originated in missouri in the U.S.A. The size of this breed is smaller than the other breeds.

Poultry System

- Zoologically the fowl belong to the genus Gallus of the family Phasianiae. The domestic fowl is called simply *Gallus domesticus.*
- Fowls are having a relatively high breathing and pulse rate, and body temperature ranges from 105°F to 109°F.
- There are about 13 common colours observed in poultry birds. These are:

(i) White (ii) Light yellow
(iii) Blue (iv) Black
(v) Barred (vi) Red
(vii) Spanglad (viii) Mottled
(ix) Laced (x) Crecentic penciled
(xi) Parallel penciled (xii) Stippled
(xiii) Striped

Division of Fowls

- Fowl may be classified on the basis of utility economic value or fency purpose and there include:

(i) Meat type (ii) Egg type
(iii) Dual purpose (iv) Game type
(v) Ornamental type (vi) Bantam

American Class

(a) Rhode Island Reds: This breed of fowl was originated from Rhode Island in New England.

- This breed is some long, rectangle body. The back is flat and the breast is carried well forward.
- The usual colour of this breed is brownish red and the wing when spread shows black both in primaries and secondaries. The tail coverts, sickle eathnrs and main tai feathers are also back. In the lower neck feather of the female, There is also slight black marking at the base.

- There are two varieties of this breed:
  (i) Single comb (ii) Rose comb
- But single comb is more popular

Plymouthrock

- This breed is most popular in America. They have single comb. Mature birds weigh from 3.5 to 4.5 kg. There are 7 varieties of plymouth rocks:
  (i) Barred (ii) White
  (iii) Buff (iv) Blue
  (v) Partridge (vi) Silver pencilled
  (vii) Columbian
- In general the plumage is greyish white each eather crossed by almost black bars which should be even in width, straight and should extend down to the skin.
- Solid black or partial black feathers. Black spots on the shanks are also common, particularly in females.

New Hampshire

- The shape of this breed body is less rectangular than Rhode island red. It has a single comb. The main tail feathers are black. In both sexes the under colours in light salmon.

  The breed is a good producer of large brown shelled eggs.

Breed of Asiatic Class

Brahma

- This breed was originated in India and exported to America and England. They have pea comb. Mature birds weight from 4 to 5 kg.
- Three varieties of Brahmas have been produced:
  (i) Light, (ii) Dark,
  (iii) Buff Brahma

- The light Brahma is most popular.

Breed of English Class

Sussex

- The breed is developed in England. It has a long body, shoulders are broad with good depth from front to rear. The males of this breed have a single comb and coloured beaks, shanks and toes.
- Two varieties of this breed given below:

  (i) Light Sussex, (ii) Red sussex

Australorp

- This breed was originated in Australia. It is a good dual purpose breed.
- The back is some what long and the body slopes gradually toward the tail. The comb is single the body is black.
- The "Austro white" a hybrid cross between the Australorp male and the white leghorn female.

Orpington

- This breed was developed in England. They are long, deep and broad and well rounded, with a full breast and a broad back. Orpingtons have single comb. Mature birds weigh from 4.5 kg.
- There are four varieties:

  (i) Buff, (ii) Black,

  (iii) White, (iv) Blue

  - Only the Buff orpington has made much popularity in America.

Breed of Mediterranean Class

Leghorn

- The breed was originated in Italy Leghorn is most popular breed. It is the world number one egg producer.

- Only three varieties are popular. They are:
  - (i) Single comb white
  - (ii) Single comb buff
  - (iii) Single comb light brown
- The breed is small, active and reputed for the harmony of its various parts. The shap of comb is quite important to leghorn fanciers.
- Leghorns are known for their stylish carriage. Mature birds weight from 2.0 to 2.7 kg.

Minorca

- Minorcas, originally called red faced black spanish, are the largest and heaviest of mediterranean breeds of poultry. Long strong bodies, large combs long wattles, large white ear lobes, The beak, shanks and toes are black.
- An excellent producer of large white eggs. Colour of skin white. The egg shell is chalk white in colour.

Various Diseases

1. Ranikhet
2. Fowl pox
3. Marek's disease
4. Fowl sporaketosis
5. Fowl cholera
6. IBD (Infectious Burasl Disease)

- The important points for consideration of poultry feed formulation

1. Feeds must contain all essential nutrients in right amount and proportion required for the purpose for which it is fed.
2. Ingredients chosen for preparation of poultry mashes must be palatable.

3. Chickens of different ages require different level of nutrients, hence only the accepted standards as per age should be followed accordingly.
4. While selecting ingredients for preparation of poultry mashes, nutritional value of each ingredient should be evaluated.
5. Chickens have no teeth to grind grains or oil cakes, hence these ingredients should be crushed into proper sizes in keeping with age of the chicken.
6. Micro-nutrients and non-nutrient feed additives should be carefully chosen and mixed up meticuously for effective results.
7. Include agro-industrial by products to minimise cost and select a variety of ingradients to make good deficiency of one by the other.
8. While selecting on ingredient care should be exercised to judge its optimum level of inclusion an many of the ingredients are likely to be deleterious at higher level.
9. Fungal infested ingredients should always be avoided.

Goat Breed

- Goats which belongs to the genus Capra has possible been developed from the following 5 wild species
    1. Capra hircus (Bezoar)
    2. Capra ibex (ibexes)
    3. Capra caucasica
    4. Capra pyrenaica
    5. Capra falconeri
- The group of goats is called Flock or Band.
- Adult male called Buck.
- Adult female known as Doe.
- Young male is known as buckling.
- Young female is known as goatling.

- New born one of goat is called kid.
- Act of parturition — kidding.
- The goat milk is easily digestible because of smaller sized fat globules making softer curd.
- The mineral content is slightly higher than cow due to presence of higher amount of calcium, phosphorus and chlorine content.

Goat meat is commonly called Chevon.

Indian Goats Breed

- Recently there are 20 breeds of goat in India. Although majority of them do not have specified defined characters.

The breeds are now classified on the basis of four agro-climatic condition of the country.

1. The North western and central arid and semi-arid region
2. The Southern peninsular region
3. The Eastern region
4. Northern temperate region

The majority of the breeds (60% of the total) are found in the north-western region.

Breed of North-western Arid and Semi-arid Region

- This region has the largest number of goats. Punjab, Haryana, Rajasthan, Gujarat, Plains of U.P. and M.P. include in this region.

Jamunapari

- The origin of this breed is between Jamuna, Gangas and Chambal rivers. It is the biggest and most popular breed of India. The breed has been extensively utilised for milk and meat (dual purpose).
- The ears are very long, flat and drooping. Both sexes are horned with short and thin tail, a thick

growth of hair is present on the buttocks, known as feathers.

- The milk production of goat is 1.5 to 2.0 kg per day.
- Usually doe kids once a year.

Barbari

- Probabily the originated of this breed was in the city of Barabera in British Somali land in east Africa.
- Small animals with compact body. The colour of this breed is white with small light brown patches. Ears are short, tubular and erect.
- Daily milk yield averages about 750 ml to 1000 ml.
- It may kid twice in a period of 12-15 months.

Beetal

- The breed is found throughout the state of Punjab and Haryana.
- The body weigh of adult male is 50-62 kg and female weight is 35-40 kg.
- The breed is good dairy type, the ears are long and flat, curled and drooping. Both sexes have thick, medium sized horns. The tail is small and thin.
- The average daily yield is 2.0 kg and per lactation varies from 150-190 kg.

Surti

- The breed is good dairy breed. The colour is white. Widely distributed in Surat and Baroda, Body weigh of male 30 kg female is 32 kg.
- Medium sized breed, ears are medium in size. The breed is unable to walk long distances. They are most economic. Both sexes have small horns directed backward.
- The breed is good milk producer, the average yield is 2.0 kg/day.

Marwari

- The breed found in the region of Rajasthan.
- The body weight of male is 30 kg and female is 25 kg.
- The breed is medium size. The male has a thick bread but this is absent in females.
- The ears are small and flat carried on a small head. Tail is small and thin, udders are fairly well developed.
- The breed is poor milker, about 0.8 kg/day average milk production but good for meat.

Mehsana

- The breed are mainly found at Mehsana, Gandhi Nagar, Ahmedabad district of Gujarat. The breed is utilized for dual purpose (for meat and milk).
- Body weigh of male is 36 kg and female is 32 kg.
- The Mehsana breed is large sized. Their colour is black with white spots at the base of ear. Ears are always white.
- The milk averages yield 1.0 kg/ day.

Jhakrana

- The breed is found in Jhakrana and few surrounding village near Behror of Alwar district of Rajasthan.
- This breed is very good dairy type.
- The body weight of adult male is 55 kg and adult female weight is 45 kg.
- The breed is large and colour is black with white spots on ears and muzzle.
- These breed are used mainly for milk production. The average milk yield varies from 2.0-3.0 kg daily.
- Kidding is mostly single but in 40% cases twins are given birth.

Kutchi

- The breed is found in kutch district in Gujarat. The body weight of adult male 43 kg and female is 35 kg.
- The average milk of the breed varies around 1.5 kg daily.

Sirohi

- The breed is found in Sirohi district of Rajasthan.
- The ratio of body weight of male and female is 50 : 40.
- The medium sized animals. The colour of this breed is brown with light brown patches. The body is covered. Ears are flat and leaf like structure.
- The breed is used usually for meat. Milk yield is poor, about 0.5 kg/day.

Zalawadi:

- The breed is found in Gujarat (Rajkot and Surendra Nagar district).
- Animal are large, black colour with white marking on ear. Ears are usually long, wide leaf like structure.
- The average daily yield is 1.5 kg.

Breed of Southern Peninsular Region

Osmanabadi

- The goats are large in size. Goats colour is mostly black and rest are white brown or spotted.
- The breed are used for dual purpose for milk and meat.
- Average daily yield varies from 0.5 to 1.5 kg.

Malbari:

- The breed are mostly found in Malapuram districts of Kerala.
- The average body weight of male and female is 38 kg and 30 kg.

- The medium sized animals. The colour of this breed is fully white or black.
- All males and few females are bearded and medium sized ears.

Sangamneri

- The breed is mostly found in Poona and Ahmednagar district of Maharashtra. The body weight of the male is 38 kg and female weight is 29 kg.
- The breed is medium size. The colour is white, black or brown with spots of other colours. Ears are drooping.
- The average milk production is 0.5 to 1.0 kg per day.

Breed of Eastern Region

- Only 25% of total goat population of the country are found in this region.

Bengal

- The breed is short-legged, compact animal.
- The breed is known for excellent mutton and skin quality. Animals are small. The colour is black, grey and white also found. Beards are found in both sexes.
- The breed is very poor in milk yield. Its meat is very tender and good test.

Ganjam

- Southern Orissa is the home of this breed.
- Average body weight of male-44 kg and female-31.0 kg
- Animals are tall. Both sexes have straight horns directly upward and slightly backward.

Breed of Northern Temperate Region

- Only 2.8% of total goat population of the country found in this region.

Changthangi

- Largely the breed are found in changthang region of Laddakh. The breed is also known as Pashmina.
- Medium sized animals. Some animals are white in colour and some are black, grey or brown colour.
- This breed looks pretty and are used for transport in hilly areas. Body is covered with long coarse hair.
- Apart from meat, the hair is used for making of shawl (Kashmiri shawl).

Gaddi

- The breed also known as Himalayan breed because this breed mostly found in hilly tract.
- The average body weight of males is 27 kg and female weight is 25 kg.
- The animals are medium sized. The colour is mostly white but black and brown also seen. Both sexes have large horns, ears are medium, long and drooping skin is very tough.

Division of Goat Breeds according to Functions

Dairy/ Meat purpose	*Meat purpose*	*Fibre purpose*
Beetal	Bengal	Gaddi
Jhakrana	Ganjan	Marwari
Jamunapari	Malabari	Chigu
Barbari	Kannai Adu	Changtihani
Surti		Kutchi
	Sangamneri	
	Osmanabad	
	Zalawadi	
	Gohilwadi	
	Mehsana	
	Sirohi	

- Goat's Common Disease

(i) Mastitis (ii) Foot-rot
(iii) Brucellosis (iv) Internal parasites
(v) External parasites (vi) Poisoning
(vii) Bloat

Sheep

- The domestic sheep belong to the phylum Chordata, class -Mammalia, family-Bovidae.

Common Aspects

- The species of sheep is called as "Ovine".
- The group of sheeps are known as "Flock".
- Adult male of sheep is "Ram" and adult female is called as "Ewe".
- New born of sheep is known as "Lamb".
- The female with its offspring is called as "Suckling".

Sheep's Breed

- Approximately over 800 breeds of sheep found in the world in the various sizes, shapes, types and colours.
- Widely in India four main types of sheep are found.

(i) The temperate Himalyan region sheep.
(ii) The Northern-Western region sheep.
(iii) Southern region sheep
(iv) The eastern region sheep

Breed of Temperate Himalayan Region

In this region comes, Hilly are like-J & K, Himachal pradesh, Garhwal district of Uttaranchall.

Gaddi

- The breed is found in hilly area (Kullu and Kangra valley).
- The breed are small size, short tail and ears. They are mostly white colour with coloured face. Flock graze

high mountains during summer and are moved to lower slopes for the winter.

- The fleece is fine and lustrous and can grow to 5 inch length used for manufacture of high quality shawls and blankets.

Bhakarwal

- The breed originates from lower hills of Himalayas.
- Animals are long sized, sturdy and straight backed.
- The sheep are clipped thrice in a year.

Gurez

- The breed originated from Gurez tehsil of Kashmir state.
- The Rams are virile and active and the ewes are good milkers. The wool is good quality. Mutton is considered to be of superior quality.

Rampur Bushar

- The medium sized breed, colour is white.
- The fleece is of medium quality. Male are always horned while females are polled.

Poonchi

- The breed is found in Poonch district of Kashmir.
- Most of the sheep are hornless, tail is short and thick at the base and colour is white.
- The breed is best for wool production.

1. Kashmir Merino

- This breed has originated from crosses of different merino types and with migratory Indian breeds of Gaddi, Poonchi.

Breed of North-Western Region

Hissardale

2. Chokla: The breed is found in Rajasthan. Head of this breed is small with brown in colour.

3. Magra: The breed found in Bikaner, Churu district of Rajasthan.
4. Nali: Large size breed compact head.
5. Marwari
6. Sonadi
7. Kathiwari: The home of the breed is Kathiwar, medium sized breed.

- The wool is good quality and coarse.

8. Jaisalmeri
9. Malpura

Breed of Southern Region

1. Deccani	2. Nellore
3. Bellary	4. Mandya
5. Coimbatore	6. Mecheri
7. Madras Red	8. Vembur
9. Hassan	10. Tiruchy Black

Breed of Eastern Region

1. Chhottanagpuri	2. Ganjam
3. Tibetan	4. Shahabadi

- Nilgiri, Hissardale, Karnal, Kashmir merino breed of sheeps are better for garments wool production.
- Gaddi, Ckokla, Nali, Gurez, Poonchi etc. breed of sheep is best for carpest wool production.
- Bellary, Marwari, Malpura, Pugal, Deccani, Jalauni, Muzzafarnagar, Chhottanagpuri, Sonadi, Balangir etc. breed of sheep better for dual purpose (meat and carpet wool production).
- Hassan, Nellore, Mandya, Vembur, Madras red, Meehari, Kanguri etc. better for specially meat purpose.

8

ANALYSED RESULTS

EXPERIMENTAL RESULT ANALYSIS

Once the experimental results are obtained they have to be analysed and interpreted. In experimental situations we may have large number of treatments. Our interest will be to test whether all the treatments have the same effect. In other words, our intention will be to test the null hypothesis, $H_0 : \mu_1 = \mu_2 = \mu_3 = ... = \mu_t$, against the alternative hypothesis, $H_1 : \mu_1 \neq \mu_2 \neq \mu_3 \neq ...... \neq \mu_t$. For this purpose we may use Student's t-test by taking all possible combinations. But it is a tedious procedure and theoretically not valid. The appropriate method for such tests is the *analysis of variance*.

The analysis of variance can be regarded as a special development of regression analysis. The fundamental principles of analysis of variance and its most important techniques were developed by R.A. Fisher. In the early stages, the analysis of variance technique was primarily applied to agricultural and biological experiments. Now it is used in every field of science.

Analysis in Different Forms

The analysis of variance is the systematic algebraic procedure of decomposing the overall variation in the responses observed in an experiment into different components. Each component is attributed to an identifiable cause or source of variation. The structure of these component parts is determined by the design of experiments.

Suppose that we have a factor *T* with *t* levels (that is, *t* treatments) with equal replication, *r* and their responses, *Y*. The results may be tabulated as shown in the following table.

The Arrangement of Data

	Treatments				
	1	*2*	*i*	*t*	
	Y_{11}	Y_{21}	Y_{i1}	Y_{t1}	
	Y_{12}	Y_{22}	Y_{i2}	Y_{t2}	
	..	..	..	..	
	Y_{1j}	Y_{2j}	Y_{ij}	Y_{tj}	
	..	..	..	..	
	Y_{1r}	Y_{2R}	Y_{ir}	Y_{tr}	
Total:	$Y_{1\cdot}$	$Y_{2.}$	$Y_{i\cdot}$	$Y_{t.}$	$Y..$ = Grand Total
Mean:	Y_1	Y_2	Y_i	Y_t	

The response of the jth individual unit in the ith treatment may be represented by the equation.

$$Y_{ij} = Y_i + e_{ij} = \text{i}^{\text{th}} \text{ treatment mean + random deviation}$$

$$= \mu + (Y_i - \mu) + (Y_{ij} - Y_i)$$

or, $$(Y_{ij} - \mu) = (Y_i - \mu) + (Y_{ij} - Y_i)$$

That is, the variation of an individual response from the overall mean is the sum of two components. One is the variation in the means of the treatments, Y_i from the overall mean, μ.

The other is the variation within each of the treatments. The variation within each treatment is not accounted for by the treatment, and hence it is the error term.

The significance of the difference $(Y_i - \mu)$ is based on the magnitude of the components $(Y_i - \mu)$ compared with $(Y_{ij} - Y_i)$. The statistical test of the significance involves the use of the following identity.

$$\Sigma\Sigma(Y_{ij} - \mu)^2 = \Sigma\Sigma\left[(Y_i \div \mu) + (Y_{ij} - Y_i)\right]^2$$

$$= \Sigma\Sigma(Y_i - \mu)^2 + \Sigma\Sigma(Y_{ij} - Y_i)^2 + 2\Sigma\Sigma(Y_i - \mu)(Y_{ij} - Y_i)$$

$$= \Sigma\Sigma(Y_i - \mu)^2 + \Sigma\Sigma(Y_{ij} - Y_i)^2$$

The last term will vanish because $\Sigma(Y_{ij} - Y_i) = 0$

Thus we have,

Total sum of squares = treatment sum of squares + error sum of square,

This can be abbreviated as,

SS(Y) = *SS(T)* + *SS(E)*

The total variation can thus be partitioned into two components, i.e., variation between treatments and variation within treatments.

Before the magnitude of the two components can be compared, they must be divided by the number of degrees of freedom. For the component "between treatments", the degrees of freedom is t - 1, and that for the "within treatments" is *(tr -t)* = *(n-t)* or $t(r$ - 1). The quantity *SS(T)/t* - 1 is known as the treatment mean square and written as *MS(T)*. The quantity *SS(E)/n -t* is known as the error mean square and written as *MS(E)*. In general, treatment mean square is denoted by s_t^2 and error mean square by s_e^2. If samples were drawn randomly, s_t^2 and s_e^2 are the independent estimates of the same quantity, σ^2, population variance of *Y*. Therefore, the ratio between two estimates,

$$F = \frac{MS(T)}{MS(E)} = \frac{s_t^2}{s_e^2}$$

follows the *F*-distribution with (*t* - 1) and (n -*t*) degrees of freedom. The significance of *F* can be determined in the usual way by using the Tables of *F*. Thus, the *F*-statistic may be used to test hypotheses about the equality of the population means of the treatments, that is, to test $H_0 : \mu_1 = \mu_2 = ... = \mu_t$. If *F* is significant the treatment means will be significantly different. The analysis of variance concept can be understood by the following examples.

Example

An experiment was conducted to study the effect of four levels of nitrogen on yield of paddy. The four levels of nitrogen were,

No application of nitrogen (t_1),

50 Kg nitrogen per hectare (t_2),

100 Kg nitrogen per hectare (t_3), and

200 Kg nitrogen per hectare (t_4).

The yields of grain in kg per plot are shown in the following table.

Grain Yield of Paddy in Kg/plot

	t_1	t_2	t_3	t_4	
	52	58	62	66	
	50	52	58	69	
	56	60	65	70	
	54	57	58	67	Grand
	53	56	61	66	Total
Total	265	283	304	338	1190

Computational steps:

Step 1: Compute the correction factor (CF)

$$CF = \frac{(\text{Grand total})^2}{n} = \frac{Y_{..}^2}{n}$$

$$= \frac{1190^2}{20} = \frac{1416100}{20}$$

$$= 70805$$

Step 2: Compute the total sum of squares

$$\text{Total} \quad SS = \Sigma\left(Y_{ij}^2 - \mu\right)^2 = \Sigma Y_{ij}^2 - \frac{Y_{..}^2}{n}$$

$$= \Sigma Y_{ij}^2 - CF$$

$$= (52)^2 + (50)^2 + \ldots\ldots + (66)^2 - CF$$

$$= (2704 + 2500 + \ldots\ldots + 4356) - CF$$

$$= 71498 - 70805 = 693$$

Step 3: Compute the treatment (or group) sum of squares.

$$\text{Treatment} \quad \text{SS} = \Sigma(\text{Y}_i - \mu)^2$$

$$= \Sigma \frac{Y_{i.}^2}{r_i} - \frac{Y_{..}^2}{n}$$

$$= \frac{(265)^2}{5} + \frac{(283)^2}{5} + \frac{(304)^2}{5} + \frac{(338)^2}{5} - CF$$

$$= \frac{1}{5}\left(265^2 + 283^2 + 304^2 + 338^2\right) - CF$$

$$= \frac{1}{5}(356974) - 70805$$

$$= 71394.8 - 70805$$

$$= 589.8$$

Step 4: Compute the error sum of squares (within treatment *SS*).

Error $SS = \Sigma\left(Y_{ij} - Y_i\right)^2$

$$= \Sigma Y_{ij}^2 - \frac{\left(\Sigma Y_i\right)^2}{r_i}$$

$$= \left(52^2 + 50^2 + 56^2 + 54^2 + 53^2\right) - \frac{265^2}{5}$$

$$+ \left(58^2 + 52^2 + 60^2 + 57^2 + 56^2\right) - \frac{283^2}{5}$$

$$+ \left(62^2 + 58^2 + 65^2 + 58^2 + 61^2\right) - \frac{304^2}{5}$$

$$+ \left(66^2 + 69^2 + 70^2 + 67^2 + 66^2\right) - \frac{338^2}{5}$$

$$= \left(14065 - \frac{70225}{5}\right) + \left(16053 - \frac{80089}{5}\right)$$

$$+ \left(18518 - \frac{92416}{5}\right) + \left(22862 - \frac{114244}{5}\right)$$

$$= (14065 - 14045) + (16053 - 16017.8)$$

$$+ (18518 - 18483.2) + (22862 - 22848.8)$$

$$= 20.0 + 35.2 + 34.8 + 13.2$$

$$= 103.2$$

Alternatively,

Error SS = total SS - treatment SS

= 693.0 - 589.8 = 103.2

These results can be put in the form of a table. This table is known as analysis of variance (ANOVA) table. For our example, the ANOVA table is as follows:

Analysis of variance for the data in Table

Sources of variation	*degrees of freedom (df)*	*sum of squares (SS)*	*Mean squares (MS)*	*F*
Between treatments	(4 - 1 =) 3	589.8	196.60	30.481**
Error	(19-3=) 16	103.2	6.45	-
Total	(20-1=) 19	693.0	-	-

In general, the ANOVA table for one-way classification will be of the form as given in Table

General form of the analysis of variance table for one-way classification.

Sources	*df*	*SS*	*MS*	*F*
Between treatments	*t* -1	*SS(T)*	$\frac{SS(T)}{t-1} = s_t^2$	s_t^2 / s_e^2
Error	*n* - *t*	*SS(E)*	$\frac{SS(E)}{n-t} = s_e^2$	
Total	*n* - 1	*SS(Y)*	-	-

For our example, table value of *F* for (3, 16) *d.f.* and at 1 percent level of significance is 5.29. Since the observed *F*-value is larger than the table value, it is significant at 1 percent level. This is indicated by two asterisks. If it is significant at 5 percent level it will be indicated by one asterisk.

The columns in the ANOVA table will be same for all types of designs. However, the rows representing the sources of variation will differ according to the type of experimental design. The form of the analysis of variance table and the computational formulae for the sums of squares used in the analysis of variance are given under each design of experiment in the subsequent chapters.

Analysis, as Model

We have seen that the jth individual unit in the ith treatment may be represented by the equation

$$Y_{ij} = \mu + (Y_i - \mu) + (Y_{ij} - Y_i)$$

This equation can be written as

$$Y_{ij} = \mu + t_i + e_{ij}$$

where, $t_i = i^{th}$ treatment effect, and

e_{ij} = error term.

This means that any Y_{ij} is made up of overall mean + treatment effect +. an error term. Since the effects are added in this model, it is known as a linear additive model. It is commonly known as *analysis of variance model.* Depending on the design, the analysis of variance model will take different forms. They are given under each design in later chapters.

Analysis of Variance: Its Two Types of Models: Usually the treatments in an experiment are fixed by the experimenter. They are not selected randomly from all possible treatments. For example, in a fertilizer experiment we may select three levels of nitrogen as 60, 80 and 100 Kg per hectare and test whether all these three levels of nitrogen have the same effect on yield of a crop. These three nitrogen levels were not selected randomly from all possible levels of nitrogen. They were determined by us as per our convenience. When the treatments are fixed like this, the analysis of variance model is known as *fixed effects model* or *Model I.*

There are instances in agricultural investigations where the treatments are not fixed like this. For example, we might be interested in the effect of geographic locations on the yield of a crop. It is possible that our concern might be with certain specific locations. In that case we would employ Model I. Instead, if our interest is to generalize to all possible locations from which they have been selected we would employ model II. As another example, suppose that we wish to determine the nitrogen content in paddy plants. We may select 5 hills at random from large number of hills. From each selected hill three independent determination of the nitrogen content may be made. The results may then be generalised to all hills. Thus, when t treatments are selected at random from a population of T treatments, we have *random effects model* or *Model II.* It is also known as *variance-components model.*

Under model I, it is assumed that all treatments about which inferences are to be made are included in the experiment.

If the experiments were to be replicated, the same set of treatments would be included in each of the replication. On the other hand, under Model II, a different random sample of t treatments would be included in each of the replication.

For both model I and model II, the equation is same. For example, the equation

$$Y_{ij} = \mu + t_i + e_{ij}$$

is appropriate for both the models. However, the assumptions underlying these models are not same. In model II, the term t_i is a random variable. The distribution of t_i is assumed to be normal with mean 0 and variance σ_t^2. The variance 'between treatments' will be $\sigma^2 + \sigma_t^2$, where σ^2 is the error variance. The assumptions about the error term are, however, same in both the models.

The procedures will be same for both the models in initial significance tests. After the initial significance test, a Model I analysis of variance is completed by examining the data in greater details. In other words, the null hypothesis under model I is $H_0 : \mu_1 = \mu_2 = \ldots\ldots = \mu_t$. In model II, we will be interested in estimating the variance σ_t^2 rather than the individual t_i. The null hypothesis under model II is, therefore, $H_0 : \sigma_t^2 = 0$. The relative amounts of variation 'between treatments' $(\sigma^2 + \sigma_t^2)$ and 'within treatments' (σ^2) would guide us in designing further studies of this sort. If 'between treatments' variation is significantly more than 'within treatments' (error) variation, we would need more treatments and fewer replications per treatment. On the other hand, if 'between treatments' variation is not significantly more than 'within treatments' variation, we would need fewer treatments and more replication per treatment.

When more number of factors are used, we may have a mixed model. If all the levels of one factor are used while a random sample of the levels of another factor is used in an experiment, a mixed model results.

Model I is better understood than the other models discussed above. Hence, it is used very commonly. It is the most appropriate model for single factor experiments. However, in multi-factor experiments choice should be made between model I and model II.

Assumptions

There are certain basic assumptions underlying the analysis of variance model. They are:

1. The effects of the different factors (treatments and environment) are additive.
2. The experimental errors are independent.
3. The experimental errors are distributed normally with mean 0 and common variance σ^2.

The usual interpretation of the analysis of variance is valid only when such assumptions are met. When the data deviate from these assumptions to lesser extent the analysis of variance technique will not be affected seriously. However, larger deviations from these assumptions will affect both the level of significance and the sensitivity of *F* and *t* tests. When the level of significance is affected, the results may look more significant than they actually are. For example, the difference between the effects of two treatments may be significant only at 6 percent level. But we may declare that it is significant at 5 percent level. This will lead to invalid conclusions.

The meaning of the above assumptions, identification of failure of these assumptions and the remedial measures are described in the following sections.

The Additivity

When the effects of two factors are independent, the effects of one factor remain constant over all levels of the other factor. In that case, the effects of two factors are said to be additive. For example, consider the hypothetical data in the following table.

Numerical example for additivity.

Treatment		*Replication*			*Replication effect*		
		I	II	III	II-I	III-I	III-II
1		20	25	35	5	15	10
2		30	35	45	5	15	10
3		40	45	55	5	15	10
Treatment	2-1	10	10	10			
effect	3-1	20	20	20			
	3-2	10	10	10			

From Table, it can be seen that the effect of treatment 2 (treatment 2 -treatment 1) is 10 for ill replications. Similarly, the effect of treatment 3 is 20 (treatment 3 - treatment 1) or 10 (treatment 3 - treatment 2) for all replications. In the same way, it can be found that the effect of replication II is 5 for all the treatments and the effect of replication III is 15 (replication III - replication I) or 10 (replication III - replication II) for all the treatments. Here the effects of treatments and replications are additive.

Let us now consider the hypothetical data in the table below:

Numerical example for non-additivity.

Treatment		*Replication*			*Replication effect*(%)		
		I	II	III	II-I	III-I	III-II
1		20	30	40	50	100	33
2		30	45	60	50	100	33
3		40	60	80	50	100	33
Treatment	2-1	50	50	50			
effect (%)	3-1	100	100	100			
	3-2	33	33	33			

The percentage increases (effects) are computed as

$$\frac{30-20}{20} \times 100, \frac{45-30}{30} \times 100, \frac{60-40}{40} \times 100.$$ and so on.

From Table, it can be observed that the effect of replication II is 50 percent for all treatments. The replication effect is, therefore, not constant over all treatments. The treatment effects are also not constant for all replications. The increase is by a percentage in this case.

When the effects are by certain percentages as in the example, they are said to be *multiplicative* or *non-additive.* If the analysis of variance model for additive effects is given by

$$Y_{ij} = \mu + t_i + r_j + e_{ij}$$

the model for multiplicative effects may be given by

$$Y_{ij} = \mu + t_i + r_j + (tr)_{ij} + e_{ij}$$

For the purpose of examining additivity we may use Tukey's test for non-additivity. For details about this test one can refer Snedecor and Cochran (1967).

In agricultural experiments, the presence of non-additive effects is common in fertilizer trials and insect and disease studies. In fertilizer trials, the plots with low level of available nitrogen in soil, for example, will benefit more from addition of nitrogen than plots with adequate level of available nitrogen. In studies on the incidence of insects or diseases, the incidence pattern will be in multiples of the initial incidence. This happens because of the nature of the growth of insect or disease organisms. In these situations, the assumption of additivity would not be correct.

When non-additive effects are present, the experimental error will be inflated. Consequently, the precision will be lost. In order to overcome this problem the multiplicative effect may be corrected by applying logarithmic transformation.

The Normality

We know that many statistical tests are valid only when the data follow normal distribution. In some experimental situations the distribution of errors may not be exactly normal as assumed under the analysis of variance model. It may be skewed or it may follow a Poisson distribution or a binomial distribution. There are many methods to test the normality. For testing the normality we need fairly large samples. With small samples it will be difficult to prove normality.

When extreme values are present in a treatment it is an indication of skewed distribution. If the ratio of largest value to smallest value is 2 or more, we may say that there are extreme values.

The nature of the functional relationship between the mean and variance may help in identifying the non-normal distribution. The functional relationship may be $\sigma^2 = \mu$. In such a case, the variance is proportional to the mean. This kind of relationship exists when the distribution is Poisson in form. We can also identify Poisson distribution by the nature of data. Often, counts of events having a small probability of occurrence follow Poisson distribution. For example, count data, such as the number of insects per plant, the number of insects caught in a trap, the number of infested plants per plot, the number of weeds per plot, the number of lesions per leaf, etc., usually will approximate Poisson distribution.

The functional relationship, $\sigma^2 = \mu(1-\mu)/n$ occurs when the distribution is binomial. When the observations are proportions or percentages like germination percentage of seed materials, percentage mortality of insects, percentage of infestation and percentage of barren tillers will approximate binomial distribution. In the usual notation of binomial distribution, $\mu = p$ and $1-\mu = q$.

The non-normal distribution of errors produce too many significant results. Also, there will be loss of efficiency in the estimate of the treatment means.

In order to normalise a non-normal distribution, we can use data transformation. Logarithmic transformation is used to normalise a skewed distribution; the square root transformation is used to normalise a Poisson distribution; and the angular transformation is used to normalise a binomial distribution.

Variance Homogeneity

In the analysis of variance the usual estimate of error variance is the pooled variance. This pooling is valid only when the variances are homogeneous. That is why the assumption of common error variance is made in the analysis of variance.

The meaning of common error variance can be explained with an example. The error term for individual responses of each treatment can be written as

$e_{ij} = Y_{ij} - Y_i$

The mean and variance for these e_{ij} values can be computed for each treatment as given in table given below.

Numerical example for common variance of error term.

Treatment				e_{ij} *for treatment*			e_{ij}^2 *for treatment*		
	1	2	3	1	2	3	1	2	3
	2	7	10	-1	1	0	1	1	0
	4	8	12	1	2	2	1	4	4
	3	4	8	0	-2	-2	0	4	4
	1	5	9	-2	-1	-1	4	1	1
	5	6	11	2	0	1	4	0	1
Total	15	30	50	0	0	0	10	10	10
Mean	3	6	10	0	0	0	2	2	2

From Table, it can be seen that the mean of error term for each treatment is 0 and the variance is 2.

In practice we may not get the variances exactly equal. However, they should be close to each other. The equality of many variances may be tested by using Barlett's or the test proposed by Hartley.

Hartley's test is simpler as compared to Bartlett's test. When the number of replications is same for all the treatments, the test statistic proposed by Hartley is given by

$$F_{max} = \frac{\text{Largest variance of the } t - \text{treatment variances}}{\text{Smallest variance of } t - \text{treatment variances}}$$

Hartley has tabulated the sampling distribution of F_{max} statistic. Against the number of treatments *(t)* and the replication degrees of freedom (r - 1), the table value can be read. If the observed F_{max} value is greater than the table value for a specified level of significance, then the variances are declared as heterogeneous.

Yet another simple method for detecting the variance heterogeneity is to use the relationship between the range and the mean. The range tends to be proportional to the variance. Hence the range is used in place of variance. When the ranges and the means of treatments are plotted on a graph sheet, it will result in a scatter diagram.

Usually, heterogeneity of variance occurs when the variance is functionally related to the mean. We have seen some of the functional relationships between the variance and the mean in Section earlier. The functional relationship may also be of the forms, $\sigma^2 = C^2\mu$ and $\sigma^2 = C^2\mu^2$, where C is a constant. These types of variance heterogeneity are usually associated with non-normal distributions.

There is another type of variance heterogeneity caused by the nature of treatments. For example, plots treated with certain insecticides may have higher variation because the insect population is not uniform or due to non-uniform application of insecticides. The variances of varieties that are highly tolerant to moisture stress are expected to be small compared to that of highly susceptible varieties. The heterogeneity of variances may be there due to the genetic variability of the varieties tested. We know that the genetic variability in F_2 generation is much higher than that of the F_1 generation.

Once the variances are found to be heterogeneous, the simplest remedy is to omit the treatments with very large variance from the analysis. If there are too many treatments to be omitted, then this method is not considered satisfactory. Instead, we can use variance stabilising transformations depending on the functional relationship between the variance and the mean. Often the assumptions of homogeneous variance and normality are violated simultaneously. If the variance is not homogeneous then the distribution is not usually normal. Hence the transformations used to normalise the distribution, viz., log, square root and angular transformations, will result in variance homogeneity.

When the variances are not functionally related to the means we may resort to *error partitioning* to handle data with heterogeneous variance. This method should be applied only after eliminating gross errors, if any.

For error partitioning we have to group the treatments with homogeneous variances. For example, suppose that the treatments are paddy varieties and that they are classified as long, medium and short duration varieties having equal variances within each group. Then the sums of squares for varieties and error will be partitioned as follows:

Varieties   Between groups

Varieties within group 1

Varieties within group 2

Varieties within group 3

Error	Replication x between groups
	Replication x group 1
	Replication x group 2
	Replication x group 3

The *F*-values are then calculated as the ratio of the 'group mean square' and the corresponding 'error mean square'. For example,

$$F(\text{group 1}) = \frac{\text{Within group 1 MS}}{\text{Replication x Group 1 MS}}$$

Error's Freedom

If the error of an observation is not dependent upon that of another, it is said to be independent. Under certain conditions it is possible that the experimental errors are correlated. In field experiments, it is often found that adjacent plots give similar yields. If we allocate the same treatments to adjacent series of plots the errors tend to vary together. For example, consider the following layout:

Replication I	*A*	*B*	*C*	*D*	*E*
Replication II	*E*	*A*	*B*	*C*	*D*
Replication III	*D*	*E*	*A*	*B*	*C*

In this layout treatments *A* and *B* are adjacent to each other in all replications. The error for treatments *A* and *B* can be expected to be more related as compared to that between *A* and *D*.

Non-independence of errors tends to produce too many significant results in *F* and *t*-tests.

It is difficult to detect the presence of non-independence of errors by inspection of the data alone. However, by inspecting the layout the non-independence of errors may be detected. The remedy to overcome the lack of independence of errors is the proper randomization.

Transformation of Data

We have seen that in actual experiments the ideal conditions for the analysis of variance may not be obtained. When one or more assumptions fail we are left with two alternatives. One is that a new model can be developed to which the data may conform. The second is that certain corrections may be made in the data in such a manner that the corrected data meet the assumptions of the analysis of variance model. Development of a new model is not easy. But it is easy to make corrections in the data. The corrections are done by means of converting the data from their original form to a new form. This conversion of data is known as *transformation* of data. Different types of transformations are available. The most commonly used transformations are log transformation, square root transformation and angular transformation.

Transformation of Log

When the original observation *Y* is converted to log *Y*, the conversion is known as log transformation. Although log to any base can be used, log to base 10 is generally the easiest. If the observed value is 0, a constant value preferably 1 is added to avoid negative logarithms. When such constant is added, it is added to all the observations.

The log transformation is particularly effective in normalising positively skewed distributions. It is also used to achieve additivity. An example for log transformation is given in the table below.

Observed values and their log transformed values.

Treatment	*Original Values*			*Log Values*		
	Replication			*Replication*		
	I	*II*	*III*	*I*	*II*	*III*
1	20	30	40	1.30	1.48	1.60
2	30	45	60	1.48	1.66	1.78
3	40	60	80	1.60	1.78	1.90

It can be verified that the treatment and replication effects are not additive in case of original observations. However, the treatment and replication effects are additive after log transformation.

Transformation of Square Root

If the original observation Y is converted to a new value by taking its square root, it is known as square root transformation. It is used in case of count data. When some of the observed counts are numerically small, say less than 10, the more appropriate transformation is $\sqrt{Y+0.5}$. The transformation of the type $\sqrt{Y}+\sqrt{Y+1}$ is also used.

The square root transformation is used when the observations follow a Poisson distribution.

Transformation of Angles

In case of proportions, derived from count data, the observed proportion p can be changed to a new form $sin^{-1}\sqrt{p}$. This type of transformation is known as angular transformation. It is also known as 'arcsine' transformation or 'inverse sine' transformation.

It may be noted that the angular transformation is not applicable to percentage data which are not derived from count data. For example, percentage of marks, percentage of profit, percentage of protein in rice, infection index, etc, can not be subjected to angular transformation.

The computational work can be reduced greatly by using ready made table of angular transformation. In the tables, the values of p are given in percentages.

The angular transformation is not good when $p=\frac{o}{n}=0$ or $p=\frac{n}{n}=1$. The transformation is improved by replacing $\frac{o}{n}$ with $\frac{1}{4n}$ and $\frac{n}{n}$ with $1-\frac{1}{4n}$, where n is the total number of units under observation.

When all the observed proportions lie between 30% and 70%, the angular transformation need not be used. This is because of the reason that the transformation in that range of values will not alter the conclusions.

The angular transformation is used to normalise the binomial distribution, especially when the observed proportions are in the range of 0 to 30% or 70 to 100%.

All the above transformations are used mainly to stabilise the variances. For the sake of convenience the situations in which the various transformations are used are summarized below:

Observed distribution	*Functional relationship between variance and mean*	*Transformation*
Poisson	$\sigma^2 = \mu$	Square root: $\sqrt{Y+0.5}$
Empirical	$\sigma^2 = C^2\mu$	Square root
Binomial	$\sigma^2 = \mu(1-\mu)/n$	Angular: $sin^{-1}\sqrt{p}$
Empirical	$\sigma^2 = C^2\mu^2$	Logarithm: Log Y

In making the choice of a transformation, we have three alternative procedures. The nature of data itself may suggest the appropriate transformation. As a second method, the graph showing the relationship between the range and the mean may be used. The third alternative is to use Tukey's test of additivity.

Once the transformation has been made, the analysis is carried out with the transformed data. The conclusions are drawn from such analysis. However, while presenting the results, the means and their standard errors are transformed back into original units. While transforming back into the original units, some corrections have to be made. In case of log transformed data, if the mean value is X, the mean value in the original units will be

$$Y = \text{antilog}\left[(X + 1.15V(X)\right]$$

instead of Y = antilog X.

If the square root transformation had been used, then

$$Y = (X + V(X))^2$$

instead of $Y = \overline{X}^2$

In the above formulae $V(X)$ is the variance of the mean X.

Presentation of Anova Results

We have seen that the null hypothesis of equal treatment means, that is, $\mu_1 = \mu_2 = \ldots\ldots = \mu_t$ can be tested using the analysis of variance technique. The test statistic in this case is F which is the ratio of the treatment mean square to the error mean square. The calculated F value is then compared with the table value of F. The table value is read against the specified level of significance and treatment and error degrees of freedom.

If the calculated value of F is greater than the table value, then F is significant; otherwise F is not significant. If F is significant at 5% level of significance, it is indicated by placing one asterisk(*) on the computed F-value. If it is significant at 1% level of significance, two asterisks (**) are placed on the computed F-value.

A non-significant F may result either due to small treatment difference or a very large experimental error or both. It does not mean always that all the treatments have the same effect. When the experimental error is large, it is an indication of the failure of the experiment to detect treatment differences. In order to find out the reliability of the experiment, the coefficient of variation (CV) is used. It is computed as

$$CV = \frac{\sqrt{\text{Error } MS}}{\text{Overall mean}} \times 100.$$

If the CV is 20% or less, it is an indication of better precision of the experiment. When the CV is more than 20%, the experiment may be repeated and efforts made to reduce the experimental error.

In case of significant F, the null hypothesis is rejected. Then the problem is to know which of the treatment means are significantly different. Many test procedures are available for this purpose. The most commonly used tests are the *least significant difference (LSD)* and the *Duncan's multiple range test (DMRT)*. The least significant difference is also known as *critical difference (CD)*.

Critical Difference

The CD is a form of t test. Its formula is given by

$CD = t \cdot SE(d)$

where,

$$SE(d) = \sqrt{EMS\left(\frac{1}{r_i} + \frac{1}{r_j}\right)}$$

In case of equal replications $SE(d) = \sqrt{\frac{2EMS}{r}}$

In the formula, t is the critical (Table) value of t for a specified level of significance and error degrees of freedom; and r_i and r_j are the number of

replications for r^{th} and j^{th} treatments, respectively. This formula is derived from the formula for t to test the significance of the difference between two means:

$$t = \frac{Y_i - Y_j}{\sqrt{s^2\left(\frac{1}{r_i} + \frac{1}{r_j}\right)}}$$

Two treatments are declared significantly different at a specified level of significance if their difference exceeds the calculated CD value; otherwise they are not significantly different.

The advantage of CD is that it is easy to calculate and it provides a single value for testing all differences. However, there are certain limitations in the use of CD. It should be used only if F is significant. It should be used only if the pair comparisons are planned before the data are examined. Comparison of the control with each of the other treatment is a common example for planned pair comparison. If the comparison is selected after the treatment means are observed, a certain number of differences in a set of t treatments will be large owing to sampling variation.

Multiple Range Test of Duncan: In a set of t treatments if comparison of all possible pairs of treatment means is required we can use Duncan's multiple range test. The DMRT can be used irrespective of whether F is significant or not. It involves number of steps which are as follows:

Step 1: Arrange the treatments in decreasing order, that is, according to their ranks.

Step 2: Calculate the standard error of mean as

$$SE(Y) = \sqrt{\frac{s_e^2}{r}} = \sqrt{\frac{EMS}{r}}$$

Step 3: From statistical Tables write the significant studentized ranges (r_p) for $p = 2, 3, \ldots\ldots t$ treatments and error degrees of freedom.

Step 4: Calculate the shortest significant ranges (R_p), where

$R_p = r_p \,.\, SE\,(Y)$

Step 5: From the largest mean subtract the R_p for largest p. Declare all the means less than this value as significantly different from the largest mean. For the remaining treatments whose values are larger than the difference (Largest mean-larges R_p), compare the differences with appropriate R_p value. If two treatments are remaining compare with R_2; if three are remaining compare with R_3; and so on.

Step 6: From the second largest mean subtract the second largest R_p and compare the treatments as in step 5.

Continue this process till all treatments are covered.

Step 7: Present the results by using either the line notation or the alphabet notation to indicate which treatments are on par and which are significantly different.

Example

From an experiment conducted in Randomized Block Design, the following results were obtained: Error mean square = 0.00193; error degrees of freedom = 12; number of replications = 4; mean of treatment 1, $(\bar{T}_1) = 0.36$; $(\bar{T}_2) = 0.16$; $(\bar{T}_3) = 0.21$; $(\bar{T}_4) = 0.16$ and $(\bar{T}_5) = 0.27$.

$$SE(d) = \sqrt{\frac{2EMS}{r}} = \sqrt{\frac{2(0.00193)}{4}} = 0.0311$$

For 12 degrees of freedom and 5 percent level of significance the table value of t is 2.18. Hence,

$CD = t \,.\, SE\ (d)$

$= (2.18)\ (0.0311) = 0.0678$

The arrangement of treatments according to their ranks is

$T_1 \quad \overline{T_5 \quad T_3} \quad \overline{T_3 \quad T_2 \quad T_4}$

The difference between the means of T_1 and T_5 is 0.36 - 0.27 = 0.09. It is greater than the CD value. Hence, T_1 and T_5 are significantly different. Since other values are less than T_5 value they are also significantly different from T_1.

The difference between the means of T_5 and T_3 is 0.27 - 0.21 = 0.06. It is less than the CD value. Hence, they are not significantly different. To indicate this a connecting line (bar) is drawn on the top (or bottom) of the treatments. It can be verified that T_5 is significantly different from T_2 and T_4; and T_3, T_2 and T_4 are not significantly different.

The presentation of results as given above is known as *bar chart*. If the treatments are not significantly different they are said to be *on par*.

The DMRT can also be applied for the above data.

Treatment	Mean
T_1	0.36
T_5	0.27
T_3	0.21
T_2	0.16
T_4	0.16

$$SE(Y) = \sqrt{\frac{EMS}{r}}\sqrt{\frac{0.00193}{4}} = 0.0220$$

p	R_p	$R_p = r._pSE\ (Y)$
2	3.08	0.068
3	3.77	0.083
4	4.20	0.092
5	4.51	0.099

The largest mean - the largest R_p = 0.36 - 0.099 = 0.261. Except the mean of T_5 all others are less than 0.261. Hence, T_1 is declared as significantly different from T_3, T_2 and T_4. Now, T_1 and T_5 are remaining and their difference is

$T_1 - T_5$= 0.36 - 0.27 = 0.09

Since two treatments are remaining this value has to be compared with R_2. The difference is more than R_2 value (0.068). Hence, T_1 and T_5 are declared significantly different. The second largest mean - the second largest $R_p = T_5 - R_4$

= 0.27- 0.092= 0.178

The means of T_2 and T_4 are less than 0.178. Hence, T_5 is significantly different from T_2 and T_4.

The remaining two treatments are T_5 and T_3. Their difference is 0.27 - 0.21 = 0.06. This is less than R_2 value. Hence they are not significantly different. In other words they are on par. To indicate this a vertical line is drawn connecting them. The third largest mean-the third largest R_p = 0.21 - 0.083 = 0.127. No treatment mean is less than this value.

Three treatments are remaining now. Hence for comparing the difference between their means, R_3 value (0.083) is taken. It can be seen that $T_3 - T_2$, $T_3 - T_4$ are less than 0.083. Hence all the three are on par.

Instead of using lines alphabets can be used as follows:

T_1	*a*
T_5	*b*
T_3	*bc*
T_2	*c*
T_4	*c*

When two treatments have a letter in common they are said to be on par. If the treatments have different letters they are declared as significantly different.

Comparison of Group

In earlier Sub-section we have considered the comparison of pairs of treatments. The pair comparison is used where there are no common characteristics among the treatments. Sometimes the treatments may have common characteristics in which case the treatments are classified into meaningful groups. Each group will consist of one or more treatments, The aggregate mean of each group is then compared to that of the others. For this purpose the degrees of freedom and sum of squares for treatments are partitioned into components.

The comparison among *t* treatments has the form

$C_t = w_1T_1 + w_2T_2 + \text{..............} + w_tT_t,$

where the *w's* are the orthogonal coefficients, *T's* are the totals of the respective treatments, and $\sum w_i$- = 0. If C_1 and C_2 are two comparisons such that

$C_1 = w_{11}\ T_1 + w_{12}T_2 + w_{13}T_3 + \ldots\ldots\ldots\ldots$

$C_2 = w_{21}\ T_1 + w_{22}T_2 + w_{23}T_3 + \ldots\ldots\ldots\ldots$

they are said to be *orthogonal comparisons* if the sum of the products of their coefficients is 0. Symbolically,

$$w_{11}\ w_{21} + w_{12}w_{22} + w_{13}\ w_{23} + \ldots\ldots\ldots\ldots = 0.$$

If C_t is any comparison among t treatments, the component of the treatments sum of squares is defined by the expression

$$\frac{C_t^2}{r\left(w_1^2 + w_2^2 + \ldots\ldots\ldots + w_t^2\right)} = \frac{C_t^2}{D_t}$$

where, r is the number of replications. Any treatment sum of squares having t - 1 degrees of freedom can be divided into t - 1 orthogonal components. This procedure may be explained with an example.

Consider the fertilizer experiment with 6 treatments - 2 organic sources of nitrogen, 3 inorganic sources of nitrogen and 1 non- fertilized control. The comparison of nitrogen sources and control is given by

$$\frac{T_1 + T_2 + T_3 + T_4 + T_5}{5} - T_6$$

where, T's are the totals for the respective treatments. This comparison can be written as

$$1 . T_1 + 1 . T_2 + 1 . T_3 + 1 . T_4 + 1 . T_5 - 5T_6$$

Similarly, the comparison of organic and inorganic sources of nitrogen is given by

$$\frac{T_1 + T_2}{2} - \frac{T_3 + T_4 + T_5}{3}$$

This can be written as

$$3T_1 + 3T_2 - 2T_3 - 2T_4 - 2T_5$$

Observe that the coefficients for one group of treatments is the divisor of the other group. The signs of these coefficients are opposite so that their sum is 0. The above functions are called *linear functions.*

The treatment sum of squares and the degrees of freedom will be partitioned as follows:

Comparison	*df*	*SS*
Control *Vs* Fertilizers	1	$\frac{(T_1+T_2+T_3+T_4+T_5-5T_6)^2}{r\left(1^2+1^2+1^2+1^2+1^2+(-5)^2\right)}$
Organic *Vs* inorganic	1	$\frac{(3T_1+3T_2-2T_3-2T_4-2T_5)^2}{r\left[3^2+3^2+(-2)^2+(-2)^2+(-2)^2\right]}$

The remaining degrees of freedom are partitioned as within organic sources and within inorganic sources.

Comparison of Trends

It is often desirable to study the nature of response of the experimental units to the varying levels of treatments like rate of fertilizer application, doses of insecticide and dates of harvest. If the treatments are in equal intervals along an ordered scale like 0, 50, 100 and 150 Kg N/ha, the treatment sum of squares may be partitioned into linear, quadratic, cubic, etc., trend components. This method of partitioning provides information about the nature of relationship between treatment levels and treatment means. Trend comparison is applicable only when the treatments are quantitative. The orthogonal coefficients for trend comparisons are taken from tables of orthogonal polynomials.

The numerical examples for above types of comparisons are given under different designs of experiments.

Plot Technique Missing

For valid use of the analysis of variance technique, the basic data should satisfy certain conditions. First, the data should conform to the assumptions of analysis of variance. Secondly, the observations from all experimental units should be available. In case of failure of the first condition we resorted to transformation of data. If the second condition is not satisfied then we use missing plot technique.

Sometimes the observations from certain experimental units may be missing due to various reasons. For example, the plants in a plot may be destroyed by stray cattle or rats or flood. The plant stands in some plots may be deficient due to poor germination or due to pest attack or due to cultural operations. In experiments on animals, an animal may die before the completion of the experiment. When measurements are made outside the field, as in the case of assessment nitrogen content of plants, 1000-grain weight, leaf arèa, etc., some portion of the sample may be lost accidentally before the measurements are taken. In some cases, data may be recorded incorrectly or extreme values may be observed which have to be omitted.

The loss of data in such cases are not due to effects of treatments. Hence they are treated as missing values. If the loss of data is due to the effects of treatments then it is not a missing value. For example, in studies of germination qualities of seed materials, a poor quality seed material may not germinate at all in a plot. Another example is an insecticide trial in which a control plot may be destroyed by the insects being controlled. In such cases, the low or zero response has to be taken as such.

In case of completely randomized design loss of data from some of the experimental units will not pose any problem since the treatments can have unequal number of replications. In other designs missing data will create problems in the analysis. The problem of missing data can be tackled by the method known as missing plot technique or by using analysis of covariance technique.

The estimation of missing values together with the analysis with missing values is generally referred to as *missing plot technique*. This technique makes best use of the available data.

The missing values are estimated using the method developed by Yates. The estimated values are entered in the body of the data. The analysis of variance is then carried out as usual. However, some modifications have to be made in the analysis of variance table and the standard error formula.

The first modification is the correction of treatment sum of squares. Usually, the estimated missing value will be positively biased. Hence the treatment sum of squares is biased upward. The amount of bias in treatment sum of squares can be found out by using appropriate formula. The estimated bias has to be subtracted from the actual treatment sum of squares to get the corrected treatment sum of squares. The estimation of bias is done only in case of one missing value. If more than one missing values are there, the treatment sum of squares is corrected by some other methods. The relevant methods are given under different designs.

Secondly, the error degrees of freedom is corrected. If one value is missing one degree of freedom is subtracted from the error degrees of freedom. In general if k values are missing, k degrees of freedom are to be subtracted from the error degrees of freedom. This is because the estimated values do not contribute to the error sum of squares.

Thirdly, the standard error formula for the difference between the mean of a treatment with missing value and that of any other treatment is different from the usual formula. The appropriate standard error formulae for different designs are given at relevant places.

Data for Multi Observation

We have seen in earlier Section that in certain experimental situations, we resort to sampling the experimental unit or make repeated measurements on the same unit at different time periods. In such cases several observations or determinations are made on each experimental unit. The resultant data are called *multi-observation data.*

Effect Interpretation: There are experiments in ected covariate may also be affected by the ich experiments the covariance analysis is used the nature of the treatment effect more clearly. onsider the experiment on the effect of levels yield of paddy. Suppose that the leaffolder ken as covariate in this experiment. It is known der incidence increases with increased level of be examined whether the difference in nitrogen d remain same after adjusting for leaffolder

1 the variate and coveriate are affected by the e interpretation of adjusted means is usually . The interpretation of results in such cases skill and experience.

bservation Data Analysis: In earlier Section it d that the covariance technique can be applied s of data with missing observations. It can now 1 some detail. Suppose one observation is missing The number of missing observation is taken as is the covariate is assigned the value 1 in the unit with missing observation and 0 elsewhere. the missing observations is taken as zero. The variance is then carried out in the usual way.

re two missing observations there will be two y X_1 and X_2. The covariate X_1 will be assigned the l as in case of one missing observation. Similarly, l the values 0 and 1. If there are more than two vations we can take as many covariates as there observations.

For the usual analysis of variance there should be a single observation per character per experimental unit. Hence we can not use the analysis of variance directly in case of multi-observation data. We may, however, use it to the average of all samples from an experimental unit or to the average of all measurements made over a time period for each experimental unit.

When multi-observation data are obtained from sampling, an additional source of variation, is introduced. It is commonly referred to as *sampling error.* Thus, in the direct analysis of multi-observation data two error terms are present. One is the experimental error and the other is the sampling error. The experimental error mean square may be significantly less or more than the sampling error mean square. In any case the experimental error mean square is used for testing the significance of the difference among treatment means. If the two error mean squares are not significantly different some researches prefer to pool them to get a single error mean square. The pooled error mean square is given by

$$\text{Pooled } EMS = \frac{\text{Experimental error } SS + \text{ sampling error } SS}{\text{Experimental error } d.f. + \text{sampling error } d.f.}$$

This pooled error mean square is then used to test the hypothesis concerning the treatment means.

In case of multi-observation data obtained from measurements over time, the analysis of variance is carried out by considering time of observation as another factor in the experiment. If the basic design is randomized block design the time of observation is treated as sub-plot treatment. The analysis will be similar to that for split - plot design. When the data is based on split - plot design, the analysis will be similar to that for split - split - plot design.

Co-variance Analysis

We have seen that the methods of grouping of experimental units into homogeneous blocks, increasing the number of replications and increasing the accuracy of the measurements are adopted to reduce the experimental error. Analysis of covariance is another method used with the same objective. The analysis of covariance is an indirect method of error control while the other methods are direct ones.

The covariance technique is used to reduce the experimental error over and above the reduction obtained by other direct methods. For the use of covariance analysis, measurements are taken on one or more additional variables that are related to the variable of primary interest. The additional related variable is known as *covariate.* In yield trials, the plant stand, the pest incidence, the quantum of the residual effects of the previous treatments, etc., are taken as covariates. In experiments on soil, the properties before the application of the treatments are taken as covariates.

The analysis of covariance is a synthesis of the methods of the analysis of variance and those of regression. The purpose of making measurements on the covariate is to adjust the measurements on the variable under study. The appropriate form of the adjustment is usually determined from the knowledge about the inter-relationship between the variables. For practical purposes, a linear relationship between the variables is presumed. If the variate is denoted by Y and the covariate by X, the adjusted mean is given by

$$Y_i' = Y_i - b(X_i - X)$$

Here, b is the regression on coefficient, X is the overall mean of the covariate, and X_i is the mean of the ith covariate. The analysis of covariance model for a observation on j^{th} unit of ith treatment in a completely randomized design has the form,

$$Y_{ij} = \mu + t_i + b\left(X_{ij} - X\right) + e_{ij}$$

The usual assumptions under the analysis of variance along with the assumptions under regression analysis are to be satisfied in the analysis of covariance model. In addition, an

assumption with r
regression effect is i

Although the cov
number of covariates
between the variate a
to one covariate whc
study is linear.

The analysis of
agricultural research i
error, (2) to assist in th
(3) to analyse the
observations.

Error Reduction

The most common
the experimental error.
Y_i is adjusted by the qu
are not directly relate
precise information on
by adjusting the Y_i for t
that the analysis of cova
If the covariate is affect
process will remove par
to highly misleading ir
that the covariate is not af
of covariance is used to
measurements are taken
of the treatments, it is obv
by the treatments. Hov
covariate are taken after
is possible that the covar
This can be tested by use c

Another point to be r
technique is that the re
significant. If b is not sign
mean will be ineffective.

Treatment
which the sel
treatment. In s
to understand
For example,
of nitrogen o
incidence is ta
that the leaffo
nitrogen. It ca
effects on yie
incidence.

When bot
treatments, th
more difficul
requires more

Missing (
was mention
for the analys
be discussed i
in any design
covariate. Th
experimental
The value of
analysis of c

If there a
covariates, sa
values 0 and
X_2 is assigne
missing obse
are missing

9

STATISTICAL DATA

THE RANDAMIZATION

We have seen that for the application of completely randomized design the experimental material should be homogeneous. Usually the experimental materials are not so homogeneous as required. It is particularly so in agricultural field experiments. In such situations the principle of local control is adopted and the experimental material is grouped into homogeneous sub-groups. The subgroup is commonly termed as *block*. Since each block will consist the entire set of treatments a block is equivalent to a replication. Hence, the terms block and replication are used synonymously in case of complete block designs.

The blocks are formed with units having common characteristics which may influence the response under study. In agricultural field experiments the soil fertility is an important character that influences the crop responses. The uniformity trial is used to identify the soil fertility of a field. If the fertility gradient is found to run in one direction, (say, from north to south) then the blocks are formed in the opposite direction (from east to west). In some experiments certain operations like spraying of insecticides may not be completed for the whole experiment in one day. Further, climatic factors like rainfall may influence the effects of the insecticides. Hence a set of treatments may be applied on the same day. In such cases a day forms a block. In animal experiments animals of same age or litter or weight may form blocks.

Analysis of Randomized Complete Block Design: Suppose the experimental material is divided into *r* blocks. Let there be *t* treatments. Each block is then divided into *t* units and the treatments are allocated within a block at random. The resulting design is called randomized complete block design. It is commonly known as *randomized block design (RBD).*

Significant Points

The principal advantage of RBD is that it increases the precision of the experiment. This is due to the reduction of experimental error by adoption of local control.

The amount of information got in RBD is more as compared to CRD. Hence, RBD is more efficient than CRD.

Flexibility is another advantage of RBD. Any number of replications can be included in RBD. If large number of homogeneous units are available, large number of treatments can be included in this design.

The statistical analysis is simple and easy. Even when some observations are missing for certain treatments, the data can be analysed by the use of missing-plot technique.

When the number of treatments is increased, the block size will increase. If the block size is large it may be difficult to maintain homogeneity within blocks. Consequently, the experimental error will be increased. Hence, RBD may not be suitable for large number of treatments. But for this disadvantage, the RBD is a versatile design. It is the most frequently used design in agricultural experiments.

For the usual analysis of variance there should be a single observation per character per experimental unit. Hence we can not use the analysis of variance directly in case of multi-observation data. We may, however, use it to the average of all samples from an experimental unit or to the average of all measurements made over a time period for each experimental unit.

When multi-observation data are obtained from sampling, an additional source of variation, is introduced. It is commonly referred to as *sampling error*. Thus, in the direct analysis of multi-observation data two error terms are present. One is the experimental error and the other is the sampling error. The experimental error mean square may be significantly less or more than the sampling error mean square. In any case the experimental error mean square is used for testing the significance of the difference among treatment means. If the two error mean squares are not significantly different some researches prefer to pool them to get a single error mean square. The pooled error mean square is given by

$$\text{Pooled } EMS = \frac{\text{Experimental error } SS + \text{ sampling error } SS}{\text{Experimental error } d.f. + \text{sampling error } d.f.}$$

This pooled error mean square is then used to test the hypothesis concerning the treatment means.

In case of multi-observation data obtained from measurements over time, the analysis of variance is carried out by considering time of observation as another factor in the experiment. If the basic design is randomized block design the time of observation is treated as sub-plot treatment. The analysis will be similar to that for split - plot design. When the data is based on split - plot design, the analysis will be similar to that for split - split - plot design.

Co-variance Analysis

We have seen that the methods of grouping of experimental units into homogeneous blocks, increasing the number of replications and increasing the accuracy of the measurements are adopted to reduce the experimental error. Analysis of covariance is another method used with the same objective. The analysis of covariance is an indirect method of error control while the other methods are direct ones.

The covariance technique is used to reduce the experimental error over and above the reduction obtained by other direct methods. For the use of covariance analysis, measurements are taken on one or more additional variables that are related to the variable of primary interest. The additional related variable is known as *covariate.* In yield trials, the plant stand, the pest incidence, the quantum of the residual effects of the previous treatments, etc., are taken as covariates. In experiments on soil, the properties before the application of the treatments are taken as covariates.

The analysis of covariance is a synthesis of the methods of the analysis of variance and those of regression. The purpose of making measurements on the covariate is to adjust the measurements on the variable under study. The appropriate form of the adjustment is usually determined from the knowledge about the inter-relationship between the variables. For practical purposes, a linear relationship between the variables is presumed. If the variate is denoted by Y and the covariate by X, the adjusted mean is given by

$$Y_i' = Y_i - b(X_i - X)$$

Here, b is the regression on coefficient, X is the overall mean of the covariate, and X_i is the mean of the ith covariate. The analysis of covariance model for a observation on j^{th} unit of ith treatment in a completely randomized design has the form,

$$Y_{ij} = \mu + t_i + b\left(X_{ij} - X\right) + e_{ij}$$

The usual assumptions under the analysis of variance along with the assumptions under regression analysis are to be satisfied in the analysis of covariance model. In addition, an assumption with respect to additivity of treatment and regression effect is implied.

Although the covariance technique can be used with any number of covariates and to any type of functional relationship between the variate and covariate, we restrict our discussions to one covariate whose relationship with the variable under study is linear.

The analysis of covariance has number of uses. In agricultural research it is used, (1) to reduce the experimental error, (2) to assist in the interpretation of treatment effects, and (3) to analyse the experimental results with missing observations.

Error Reduction

The most common use of covariance technique is to reduce the experimental error. We have seen that the treatment mean, Y_i is adjusted by the quantity $b(X_i - X)$. If the differences in X_i are not directly related to the treatment effects then more precise information on the treatment effects may be obtained by adjusting the Y_i for the variations in the X_i. It is in this way that the analysis of covariance controls the experimental error. If the covariate is affected by the treatments, the adjustment process will remove part of the treatment effect. This may lead to highly misleading interpretations. Hence, it is necessary that the covariate is not affected by the treatments when analysis of covariance is used to reduce the experimental error. When measurements are taken on the covariate before the application of the treatments, it is obvious that the covariate is not affected by the treatments. However, if the measurements on the covariate are taken after the application of the treatments, it is possible that the covariate gets affected by the treatments. This can be tested by use of analysis of variance of the covariate.

Another point to be remembered in the use of covariance technique is that the regression co-efficient, *b* should be significant. If *b* is not significant the adjustment for treatment mean will be ineffective.

Treatment Effect Interpretation: There are experiments in which the selected covariate may also be affected by the treatment. In such experiments the covariance analysis is used to understand the nature of the treatment effect more clearly. For example, consider the experiment on the effect of levels of nitrogen on yield of paddy. Suppose that the leaffolder incidence is taken as covariate in this experiment. It is known that the leaffolder incidence increases with increased level of nitrogen. It can be examined whether the difference in nitrogen effects on yield remain same after adjusting for leaffolder incidence.

When both the variate and coveriate are affected by the treatments, the interpretation of adjusted means is usually more difficult. The interpretation of results in such cases requires more skill and experience.

Missing Observation Data Analysis: In earlier Section it was mentioned that the covariance technique can be applied for the analysis of data with missing observations. It can now be discussed in some detail. Suppose one observation is missing in any design. The number of missing observation is taken as covariate. Thus the covariate is assigned the value 1 in the experimental unit with missing observation and 0 elsewhere. The value of the missing observations is taken as zero. The analysis of covariance is then carried out in the usual way.

If there are two missing observations there will be two covariates, say X_1 and X_2. The covariate X_1 will be assigned the values 0 and 1 as in case of one missing observation. Similarly, X_2 is assigned the values 0 and 1. If there are more than two missing observations we can take as many covariates as there are missing observations.

9

STATISTICAL DATA

THE RANDAMIZATION

We have seen that for the application of completely randomized design the experimental material should be homogeneous. Usually the experimental materials are not so homogeneous as required. It is particularly so in agricultural field experiments. In such situations the principle of local control is adopted and the experimental material is grouped into homogeneous sub-groups. The subgroup is commonly termed as *block*. Since each block will consist the entire set of treatments a block is equivalent to a replication. Hence, the terms block and replication are used synonymously in case of complete block designs.

The blocks are formed with units having common characteristics which may influence the response under study. In agricultural field experiments the soil fertility is an important character that influences the crop responses. The uniformity trial is used to identify the soil fertility of a field. If the fertility gradient is found to run in one direction, (say, from north to south) then the blocks are formed in the opposite direction (from east to west). In some experiments certain operations like spraying of insecticides may not be completed for the whole experiment in one day. Further, climatic factors like rainfall may influence the effects of the insecticides. Hence a set of treatments may be applied on the same day. In such cases a day forms a block. In animal experiments animals of same age or litter or weight may form blocks.

Analysis of Randomized Complete Block Design: Suppose the experimental material is divided into r blocks. Let there be t treatments. Each block is then divided into t units and the treatments are allocated within a block at random. The resulting design is called randomized complete block design. It is commonly known as *randomized block design (RBD).*

SIGNIFICANT POINTS

The principal advantage of RBD is that it increases the precision of the experiment. This is due to the reduction of experimental error by adoption of local control.

The amount of information got in RBD is more as compared to CRD. Hence, RBD is more efficient than CRD.

Flexibility is another advantage of RBD. Any number of replications can be included in RBD. If large number of homogeneous units are available, large number of treatments can be included in this design.

The statistical analysis is simple and easy. Even when some observations are missing for certain treatments, the data can be analysed by the use of missing-plot technique.

When the number of treatments is increased, the block size will increase. If the block size is large it may be difficult to maintain homogeneity within blocks. Consequently, the experimental error will be increased. Hence, RBD may not be suitable for large number of treatments. But for this disadvantage, the RBD is a versatile design. It is the most frequently used design in agricultural experiments.

RBD Layout

The entire experimental material is divided into number of homogeneous blocks. If there are *t* treatments each block is divided into *t* units. The units in each block are numbered from 1 to *t*. The treatments are also numbered conveniently.

By using random number tables, we select *t* distinct random numbers from 1 to *t*. These random numbers correspond to the treatment numbers. The first selected treatment is allotted to the first unit of a block, the second selected treatment to the second units, and so on. The randomization is done afresh for each block in the same way.

RBD Result Interpretation

The results from RBD can be arranged in two-way tables according to the replications (blocks) and treatments. There will be *rt* observations in total. The data arrangement will be as in earlier table

Arrangement of data from RBD.

Treatment	*Replication*					*Total*
	1	2	3	....	r	
1.	Y_{11}	Y_{12}	Y_{13}	....	Y_{1r}	T_1
2.	Y_{21}	Y_{22}	Y23	....	Y_{2r}	T_2
3.	Y_{31}	Y_{32}	Y_{33}	....	Y_{3r}	T_3
-	-	-	-	....	-	
t	Y_{t1}	Y_{t2}	y_{t3}	....	Y_{tr}	T_t
Total	R_1	R_2	R_3	....	R_t	G.T.

The analysis of variance model for RBD is given by

$Y_{ij} = \mu + t_i + r_j + e_{ij}$

where, μ = the overall mean,

t_i = the ith treatment effect,

r_j = the jth replication effect, and

e_{ij} = the error term.

The total variance is thus divided into three sources of variation, *viz.*, between replications, between treatments and error. The required sums of squares are obtained as follows:

$$CF = \frac{(GT)^2}{rt}$$

$$\text{Total } SS = \sum Y_{ij}^{\,2} - CF$$

$$\text{Replication } SS = \frac{1}{t}\sum R_j^2 - CF$$

$$\text{Treatment } SS = \frac{1}{r}\sum T_i^2 - CF$$

Error SS = Total SS - Replication SS - treatment SS

With these results the analysis of variance table is completed. The form of ANOVA table for RBD with t treatments and r replications each is given in Table.

Analysis of variance for RBD.

Sources of Variation	*df*	*SS*	*MS*	*F*
Replication	r - 1	*RSS*	*RMS*	*RMS /EMS*
Treatment	t- 1	*TSS*	*TMS*	*TMS /EMS*
Error	$(r-1)(t-1)$	*ESS*	*EMS*	
Total	rt-1	Total *SS*		

If *F*-test shows that there is no significant difference between replications it is an indication that the RBD will not contribute to precision in detecting treatment differences. In such situations the adoption of RBD in preference to CRD is not advantageous.

If by *F*-test we find significant difference between treatments, then we can use critical difference (*CD*) for comparing pairs of treatments.

Such differences can also be tested by Duncan's multiple range test. The *CD* is given by:

$$CD = t \cdot SE(d)$$

where, t = table value of t for a specified level of significance and error degrees of freedom.

$$SE(d) = \sqrt{\frac{2EMS}{r}}$$

Based on the *CD* value the bar chart can be drawn. Using the bar chart conclusions can be written.

Example

An experiment was conducted in RBD to study the comparative performance of fodder sorghum under rainfed conditions.

The rearranged data are given in Table:

Green matter yield of sorghum, Kg/plot.

Variety	*Replication*				*Total*	*Mean*
	I	*II*	*III*	*IV*		
African Tall	22.9	25.9	39.1	33.9	121.8	30.4
Co-11	29.5	30.4	35.3	29.6	124.8	31.2
FS-1	28.8	24.4	32.1	28.6	113.9	28.5
K-7	47.0	40.9	42.8	32.1	162.8	40.7
Co-24	28.9	20.4	21.1	31.8	102.2	25.6
Total	157.1	142.0	170.4	156.0	625.5	31.3

$$CF = \frac{(625.5)^2}{20} = 19562.5125$$

$$\text{Total } SS = \left[(22.9)^2 + (29.5)^2 + \ldots\ldots + (31.8)^2\right] - CF$$
$$= 20514.95 - 19562.5125$$
$$= 952.4375$$

$$\text{Replication } SS = \frac{1}{5}\left[(157.1)^2 + \ldots\ldots + (156.0)^2\right] - CF$$
$$= 19643.3140 - CF = 80.8015$$

$$\text{Variety } SS = \frac{1}{4}\left[(121.8)^2 + \ldots\ldots + (102.2)^2\right] - CF$$
$$= 20083.0425 - CF = 520.5300$$

$$\text{Error } SS = \text{Total}SS - \text{Rep.}SS - \text{Variety}SS$$
$$= 351.1060$$

Analysis of variance for the data in Table

Sources of Variation	*df*	*SS*	*MS*	*F*
Replication	3	80.8015	26.9338	< 1
Variety	4	520.5300	130.1325	4.448*
Error	12	351.1060	29.2588	
Total	19	952.4375		

$$SE(d) = \sqrt{\frac{2EMS}{r}} = \sqrt{\frac{2(29.2588)}{4}}$$
$$= \sqrt{14.6294} = 3.8248$$

$$CD = t\,.\,SE(d)$$
$$= (2.179)(3.8248) = 8.33$$

$$CV = \frac{s}{\mu} \times 100 = \frac{5.41}{31.28} \times 100 = 17.3\%$$

$V_4 \quad V_2$ —— V_1 —— V_3 —— V_5

From the bar chart it can be concluded that sorghum variety

K-7 produces significantly higher green matter than all other varieties. The remaining varieties are all on par.

Example

In order to assess the effects of different sources of nitrogen on paddy yield an experiment was conducted in RBD. The treatments consisted of four levels of nitrogen in the form of Ammonium chloride (A), four levels of nitrogen in the form of Urea (U) and a control (C). The net plot area was 38.31 square metres. The results are given in Table.

Grain yield of Co-33 Paddy, Kg/plot.

Treatment	Replication				Total
	I	II	III	IV	
(1) Control	3.440	3.850	4.850	4.760	16.900
(2)A N1	7.805	8.020	8.010	6.580	30.415
(3) A N2	10.040	11.325	9.980	8.670	40.015
(4)A N3	13.180	13.540	12.000	9.190	47.910
(5)A N4	13.300	13.600	13.460	14.150	54.510
			A Total		172.850
(6) U Nl	6.345	7.420	6.870	7.940	28.575
(7)U N2	10.275	10.430	10.410	10.270	41.385
(8) U N3	12.520	11.825	10.650	10.835	45.830
(9) U N4	12.985	12.975	12.950	12.760	51.670
			U Total		167.460
Total	89.890	92.985	89.180	85.155	357.210

In this case we want to compare control and sources of nitrogen, *viz.*, Ammonium chloride and Urea, nitrogen levels within Ammonium chloride and nitrogen levels within Urea. The analysis of such group comparisons is as follows:

$$CF = \frac{(357.210)^2}{36} = 3544.4162$$

Total *SS* = [(3.440)^2+ .... + (12.760)^2]- *CF*

= 322.3248

Replication *SS* = 1/9 [(89.890)^2+ .... + (85.155)^2] - *CF*

= 3.4581

Treatment *SS* = 1/4 [(16.900)^2+ (30.415)^2+ .... + (51.670)^2]- *CF*

= 300.0881

Error *SS* = 18.7786

The treatment *SS* is then split as per the required group comparisons.

Control *Vs* others *SS*

$$= \left[\frac{(16.900)^2}{4} + \frac{(172.850 + 167.460)^2}{32}\right] - CF$$

= 71.4025+ 3619.0905- *CF* =146.0768

Note that this comparison can be obtained by using the linear contrast method also. The coefficient for control is - 8 and the coefficients for others are 1, 1

Thus we have the *SS* for control *Vs* others, as

$$\frac{\left[(-8)(16.900) + (1)(30.415) + (1)(40.015) + ... + (1)(51.670)\right]^2}{4\left[(-8)^2 + 1^2 + 1^2 + 1^2 + 1^2 + 1^2 + 1^2 + 1^2 + 1^2\right]}$$

$$= \frac{(205.11)^2}{288} = 146.0768$$

A *Vs* U *SS* $= \frac{(172.850)^2}{16} + \frac{(167.460)^2}{16} - \frac{(340.310)^2}{32}$

$= 3619.9984 - 3619.0905$

$= 0.9079$

Within A *SS* $= 1/4\left[(30.415)^2 + (40.015)^2 + (47.910)^2 (54.510)^2\right]$

$- \frac{(172.850)^2}{16}$

$= 1948.2452 - 1867.3202$

$= 80.9250$

Within U *SS* $= \frac{1}{4}\left[(28.575)^2 + (41.385)^2 + (95.830)^2 + (51.670)^2\right] - \frac{(167.460)}{16}$

$= 1824.8566 - 1752.6782$

$= 72.1784$

The results are presented in the form of analysis of variance table in Table

Analysis of variance for the data in Table

Sources of Variation	*df*	*SS*	*MS*	*F*
Replication	3	3.4581	1.1527	—
Treatments	8	300.0881	37.5110	—
C *Vs* Others	1	146.0768	146.0768	186.703**
A *Vs* U	1	0.9079	0.9079	1.160
Within A	3	80.9250	26.9750	34.477**
Within U	3	72.1784	24.0595	30.751**
Error	24	18.7786	0.7824	
Total	35	322.3248		

SE(d) for C *Vs* others $= \sqrt{EMS\left(\frac{1}{4} + \frac{1}{32}\right)}$

$= 0.4691$

CD = (2.064) (0.4691) $= 0.9682$

SE (d) for levels of N within A

$$= \sqrt{EMS(1/4+1/4)}$$

$$= 0.6255$$

CD = (2.064) (0.6255) = 1.2910

SE(d) for levels of N within U

$$= \sqrt{EMS(1/4+1/4)}$$

$$= 0.6255$$

CD = (2.064) (0.6255) = 1.2910

We need bar charts only for comparing the levels of nitrogen within groups. For others the interpretations can be written directly. The bar chart for within A is

n_4 n_3 n_2 n_1

The bar chart for within U is

n_4 n_3 n_2 n_1

The conclusions can be written based on these bar charts.

If we want to present the result as yield per hectare the mean values have to be

multiplied by the conversion factor, $\frac{10,000}{38.31} = 261.03$

The *SE(d)* values have also to be multiplied by the same conversion factor. In general the conversion factor is obtained as 10,000/ plot area in square metres (since one hectare = 10000 square metres).

Example

In order to assess the effect of different insecticides in controlling the rice gall midge incidence a experiment was conducted in RBD. The result are presented in Table

Incidence of Rice Gall Midge (%).

Treatment	Replication I	II	III
Fenthion 1000 EC at 500 ml/ha (T_1)	17.8	24.4	18.1
Endosulfan 35 EC at 750 ml/ha *(T_2)*	20.2	22.7	23.0
Phosalone 35 EC at 1500 ml/ha (T_3)	15.7	16.4	18.8
Quinolphos 25 EC at 750 ml/ha (T_4)	19.7	15.3	19.8
Phosphamidon 86 WSC at 250 ml/ha (T_5)	19.4	20.3	23.7
Chlorpyriphos 20 EC at 1250 ml/ha (T_6)	17.7	18.6	21.1
Control (T_7)	21.7	21.6	17.4

The angular transformation has to be used for this data. The angular transformed values are given in Table

Angular transformed values for data in Table.

Treatment	*Replication* I	II	III	*Total*	*Mean*
T_1	24.95	29.60	25.18	79.73	26.58
T_2	26.71	28.45	28.66	83.82	27.94
T_3	23.34	23.89	25.70	72.93	24.31
T_4	26.35	23.03	26.42	75.80	25.27
T_5	26.13	26.78	29.13	82.04	27.35
T_6	24.88	25.55	27.35	77.78	25.93
T_7	27.76	27.69	24.65	80.10	26.70
Total	180.12	184.99	187.09	552.20	

The analysis of variance is carried out in the usual manner. The results are presented in Table.

Analysis of variance for the data in Table

Sources of Variation	*df*	*SS*	*MS*	*F*
Replication	2	3.6527	1.8264	< 1
Treatment	6	27.5662	4.5944	1.471
Error	12	37.4830	3.1236	
Total	20	68.7019		

Since the *F* for treatments is not significant, we can conclude that none of the insecticides have significant effect on the control of incidence of gall midge.

While reporting the results the mean value may be given in original units. For example, for treatment 1, the mean is 26.58. The mean in original units is 20.02 percent.

Example: An experiment was conducted in RBD to evaluate the effects of fungicides for the control of Thanjavur wilt disease in coconut. The results are given in the table given below.

Disease index of Thanjavur wilt disease in coconut.

Treatment	*Replication*		
	I	II	III
Kitazin 0.1%(T_1)	32.9	0	24.3
Vitavax 0.1% (T_2)	92.8	19.1	74.4
Bavistin 0.1 % (T_3)	250.9	7.8	59.6
Bavistin 0.5% (T_4)	142.6	91.5	146.5
Bordeaux mixture 1 % (T_5)	90.6	126.4	107.8
Sulphur dust 1.5 kg/palm (T_6)	19.6	16.0	185.7
Control (T_7)	172.3	161.6	98.2

The log transformation is used in this case. Since zero value is present we have to take log $(y + 1)$, instead of log y. The log transformed values are presented in Table.

Log transformed values of data in Table.

Treatment	*Replication*			*Total*
	I	II	III	
T_1	1.5302	0	1.4031	2.9333
T_2	1.9722	1.3032	1.8774	5.1528
T_3	2.4012	0.9445	1.7825	5.1282
T_4	2.1572	1.9661	2.1688	6.2921
T_5	1.9619	2.1052	2.0366	6.1037
T_6	1.3139	1.2304	2.2711	4.8154
T_7	2.2388	2.2111	1.9965	6.4464
Total	13.5754	9.7605	13.5360	36.8719

The results of the analysis of variance for the data in earlier table are presented in the following table.

Analysis of variance for the data in Table.

Sources of Variation	*d.f.*	*SS*	*MS*	*F*
Replication	2	1.3718	0.6859	3.843
Treatment	6	2.9414	0.4902	2.746
Error	12	2.1416	0.1785	
Total	20	6.4548		

Since the *F* for treatments is not significant, we conclude that the fungicides have no significant effect on the control of Thanjavur wilt disease in coconut.

Example: In a fertilizer trial conducted in RBD, the mean number of chaffs per panicle of Co-33 paddy was observed. The results are given in the table below.

Mean number of chaffs per panicle of Co-33 paddy.

Treatment	*Replication*			
	I	*II*	*III*	*IV*
Control (T_1)	9	13	12	6
A N1 (T_2)	6	9	8	14
AN 2 (T_3)	17	18	14	12
A N3 (T_4)	19	18	23	13
A N4 (T_5)	18	12	18	16
U Nl (T_6)	12	14	10	11
U N2 (T_7)	15	13	11	13
U N3 (T_8)	12	15	10	12
U N4 (T_9)	18	13	12	12

(A: Ammonium chloride; U: Urea)

The square root transformation is used for this data. The transformed value X is obtained by using the formula $\sqrt{y+0.5}$. The transformed values are give in the table below.

Square root transformed values for data in Table.

Treatment	*Replication*				*Total*	*Mean*
	I	II	III	IV		
T_1	3.08	3.67	3.54	2.55	12.84	3.21
T_2	2.55	3.08	2.92	3.81	12.36	3.09
T_3	4.18	4.30	3.81	3.54	15.83	3.96
T_4	4.42	4.30	4.85	3.67	17.24	4.31
T_5	4.30	3.54	4.30	4.06	16.20	4.05
T_6	3.54	3.81	3.24	3.39	13.98	3.50
T_7	3.94	3.67	3.39	3.67	14.67	3.67
T_8	3.54	3.94	3.24	3.54	14.26	3.56
T_9	4.30	3.67	3.54	3.54	15.05	3.76
Total	33.85	33.98	32.83	31.77	132.43	

The results of the analysis of variance are presented in the following table.

Table Analysis of variance for the data in Table.

Sources of variation	*d.f.*	*SS*	*MS*	*F*
Replication	3	0.3531	0.1177	< 1
Treatment	8	4.9368	0.6171	4.010**
Error	24	3.6929	0.1539	
Total	35	8.9828		

$$SE(d) = \sqrt{\frac{2(0.1539)}{4}} = 0.2774$$

$$CD = (2.064)(0.2774) = 0.573$$

The conclusions are drawn in the usual manner. The group comparison can be made in this case as was done in earlier Example

Example

An experiment was conducted in RBD to assess the yielding abilities of seven paddy varieties. From each replication five panicles were selected at random and the number of filled grains per panicle was observed. The results are tabulated below.

Number of filled grains per panicle of Paddy Varieties.

Variety	*Replication* I	II	III
Vaigai	120,72,151, 105,164,(612)	98,102,139, 117,107,(563)	113,99,110, 95,111,(528)
ADT-31	87,129,75 93,76,(460)	100,94,102 131,136,(563)	140,113,152 92,74,(571)

Contd...

Variety		*Replication*	
	I	II	III
TKM-9	90,68,94, 91,68,(411)	122,154,122, 83,97,(578)	73,92,109, 71,76,(421)
IR-50	69,68,96, 85,88,(406)	113,81,104, 94,89,(481)	117,79,86, 75,88,(445)
ACM-5	159,93,152, 161,128,(693)	117,124,186, 212,133,(772)	126,127,132, 116,132,(633)
Karuna	184,138,149, 130,112,(713)	148,126,128, 152,215,(769)	136,205,156, 201,102,(800)
ADT-36	100,141,101, 114,104,(560)	83,140,118, 103,137,(581)	110,110,120, 98,108,(546)

(Figures in brackets indicate the totals for the cell.)

In this problem we have multiple observations per cell. The analysis of variance procedure for this data is given below. First, replication x treatment table is prepared from earlier table as shown below.

Replication X variety table for the data in Table.

Variety		*Replication*		*Total*	*Mean*
I	II	III			
1	612	563	528	1703	113.5
2	460	563	571	1594	106.3
3	411	578	421	1410	94.0
4	406	481	445	1332	88.8
5	693	772	633	2098	139.0
6	713	769	800	2282	152.1
7	560	581	546	1687	112.5
Total	3855	4307	3944	12106	

Then other calculations are done as follows:

$$CF = \frac{(12106)^2}{105} = 1395764.15$$

From earlier table

Total SS $= \left[(120)^2 + (72)^2 + \ldots\ldots + (98)^2 + (108)^2\right] - CF$

$= 1504460 - CF = 108695.85$

From earlier table

Replication SS $= \left[\frac{(3855)^2}{35} + \frac{(4307)^2}{35} + \frac{(3944)^2}{35}\right] - CF$

$= 1399040.28 - CF$

$= 3276.13$

Variety SS $= \left[\frac{(1703)^2}{15} + \frac{(1594)^2}{15} + \ldots\ldots + \frac{(1687)^2}{15}\right] - CF$

$= 1443897.73 - CF$

$= 48133.58$

Rep. x variety

Table SS $= \left[\frac{(612)^2}{5} + \frac{(460)^2}{5} + \ldots\ldots + \frac{(546)^2}{5}\right] - CF$

$= 1453060.80 - CF$

$= 57296.65$

Error SS = Rep. X *Var*. Table SS – Replication SS - Variety SS,

=57296.65 - 3276.13 - 48133.58

= 5886.94

Sampling error SS

= Total SS - Replication SS - Variety SS - error SS,

= 51399.20

The results are presented in the table below.

Analysis of variance for the data in Table.

Sources of variation	*df*	*SS*	*MS*	*F*
Replication	2	3276.13	1638.0650	-
Varieties	6	48133.58	8022.2633	16.353**
Error	12	5886.94	490.5783	
Sampling Error	84	51399.20	611.8952	
Total	104	108695.85		

$$SE(d) = \sqrt{\frac{2EMS}{rs}} = \sqrt{\frac{2(490.5783)}{3(5)}}$$

$$= 8.0877$$

$$CD = t\,.\,SE(d)$$

$$= (2.179)(8.0877) = 17.62$$

V_6——— V_5———V_1 V_7 V_2———V_3———V_4

From the bar chart the conclusions can be written easily.

Blocking Efficiency

As already stated in earlier section if F for replications is significant, blocking is considered to be effective in reducing the experimental error. The effect of blocking can also be ascertained by finding out the relative efficiency of RBD over CRD. For this purpose the amounts of information of these designs are compared.

We know that the amount of information obtained is $\frac{1}{EMS}$. The *EMS* for CRD has to be estimated from the analysis of variance of RBD. Had the CRD been used instead of RBD, the variations due to blocks could have been added to the variations due to extraneous factors. Hence the estimate of *EMS* for CRD is obtained as the weighted average of the replication mean square and error mean square of RBD. Symbolically,

$$EMS(CRD) = \left[\frac{n_r \,.\, RMS + (n_t + n_e)EMS}{n_r + n_t + n_e}\right]$$

where, RMS = replication mean square

EMS = error mean square

n_r = replication *df*

n_t = treatment *df*

n_e = error *df*

Now, let us denote the amount of information of RBD and CRD as I(RBD) and I(CRD), respectively. The relative efficiency of RBD over CRD is obtained as

$$RE(RBD) = \left[\frac{I(RBD)}{I(CRD)}\right] \times 100$$

If the relative efficiency is more than 100 percent the excess is known as the *gain in efficiency* due to RBD.

When the error degrees of freedom is less than 20, the relative efficiency has to be adjusted by multiplying it by the *precision factor*. The precision factor is computed as

$$PF = \frac{(n_e + 1)(n_e' + 3)}{(n_e + 3)(n_e' + 1)}$$

where,

n_e = error *df* for RBD

$n_e' = n_r + n_e$ = error *df* for CRD.

Example

Consider the analysis of variance results of earlier Example.

For the sake of convenience the required information is tabulated below:

Analysis of variance results from Table

Sources of variation	*df*	*MS*
Replication	2	0.6859
Treatment	6	0.4902
Error	12	0.1785

The estimate of *EMS* for CRD

$$= \frac{[2(0.6859)] + [(6+12)(0.1785)]}{2+6+12}$$

$$= \frac{4.5848}{20} = 0.2292$$

$$I(RBD) = \frac{1}{0.1785} = 5.6022$$

$$I(CRD) = \frac{1}{0.2292} = 4.3630$$

$$RE = \frac{5.6022}{4.3630} \times 100 = 128.4\%$$

Since the error *df* is less than 20 we use the precision factor to correct the relative efficiency.

$$PF = \frac{(12+1)[(12+12)+3]}{(12+3)[(2+12)+1]}$$

$$= \frac{(13)(17)}{(15)(15)} = 0.9822$$

Crrected *RE* $= RE \times PF$

$$= (128.4)(0.9822)$$

$$= 126.1\%$$

In this case the gain in efficiency due to blocking is 26.1%

RBD Missing Plot Technique

With a single missing value in RBD with *t* treatments and

r replications each, the first step is to estimate the missing value by using the formula

$$X = \frac{rR' + tT' - G'}{(r-1)(t-1)}$$

where,

X = estimate of the missing value

R' = total of available values of the replication that contains the missing value

T' = total of available values of the treatment that contains the missing value

G' = grand total of all available values.

The computation of sums of squares is then carried out as usual. The treatment sum of squares is to be corrected by subtracting the upward bias from it. The upward bias is given by

$$B = \frac{[R' - (t-1)X]^2}{t(t-1)}$$

The degrees of freedom for error is corrected by subtracting 1 from the actual df. For the sake of completion, we may correct the total *SS* and total *df*. With these corrections the analysis of variance table is completed.

For comparing the mean of the treatment with a missing value and the mean of any other treatment, the standard error formula will be

$$SE(d) = \sqrt{\frac{EMS}{r}\left[2 + \frac{t}{(r-1)(t-1)}\right]}$$

The *SE(D)* formula for other comparisons will be the usual one.

When there are two or more missing values all but one missing value is estimated roughly as the average of the replication mean and treatment mean corresponding to the missing value. The left out missing value is estimated by using the formula for one missing value. After inserting this value, the other values are estimated one by one using the formula. The completion of estimates for all the missing values constitutes one cycle. A second approximation is then found for all the missing values in succession.

After the completion of the second cycle, the estimates from the first and the second cycles are compared. If they are equal or approximately equal the iteration process of estimation is stopped, otherwise, the process is continued. This sort of comparison is made on completion of each cycle. The estimated values in the last cycle are taken as the correct estimates of the missing values. The analysis is then carried out as usual. A separate analysis is done as CRD with the available data omitting treatment classification. In this analysis the error *SS* consists of treatment SS + error *SS*. Subtracting the error SS of RBD from error *SS* of CRD, the corrected treatment SS can be obtained. The error *df* is corrected by subtracting the number of missing values from the actual one. In case of two or more missing values one more *SE(d)* has to be computed for comparing the means of the treatments involving missing values. It is given by

$$SE(d) = \sqrt{EMS\left[\frac{1}{r_i} + \frac{1}{r_j}\right]}$$

where, r_i and r_j are the effective number of replications for i^{th} and j^{th} treatments with missing values. Suppose treatments A and *B* involve missing values and the means of these two treatments are to be compared. In a replication, the effective number of replication for *A* is counted as 1, if both A and *B* are present in the replication, as 0 if A itself is missing, and as, $\frac{(t-2)}{(t-1)}$ (*t* is the number of treatments) if only *B* is missing.

Similarly, the effective number of replications for treatment *B* is calculated.

Example: Consider the data in Table . Suppose that the value for treatment 2 is missing in replication III. The data will then be as presented in the table below:

RBD data with one missing value

Treatment	*Replication*				*Total*
	I	II	III	IV	
1	22.9	25.9	39.1	33.9	121.8
2	29.5	30.4	X	29.6	89.5
3	28.8	24.4	32.1	28.6	113.9
4	47.0	40.9	42.8	32.1	162.8
5	28.9	20.4	21.1	31.8	102.2
Total	157.1	142.0	135.1	156.0	590.2

The estimate of missing value,

$$X = \frac{rR' + tT' - G'}{(r-1)(t-1)}$$

$$= \frac{4(135.1) + 5(89.5) - 590.2}{(3)(4)}$$

$$= \frac{397.7}{12} = 33.1$$

The upward bias,

$$B = \frac{[R' - (t-1)X]^2}{t(t-1)}$$

$$= \frac{[135.1 - 4(33.1)]^2}{(5)(4)}$$

$$= \frac{7.29}{20} = 0.3645$$

After substituting the estimated missing value, we get

Treatment 2 total = 89.5 + 33.1 = 122.6,

Replication 3 total = 135.1 + 33.1 = 168.2, and

The grand total = 590.2 + 33.1 = 623.3

$$\text{Treatment } SS = \frac{1}{4}\ [(121.8)^2 + (122.6)^2 + (113.9)^2$$

$$+ (162.8)^2 + 102.2)^2] - \frac{(623.3)^2}{20}$$

$$= 19946.9725\text{-}19425.1445 = 521.8280$$

Corrected treatment SS = 521.8280 - 0.3645

= 521.4635

With these values the analysis of variance table is completed.

Analysis of variance for the data in Table

Sources of variation	*d.f.*	*SS*	*MS*	*F*
Replication	3	69.1855	23.0618	1
Treatment	4	521.4635	130.3659	4.117*
Error	11	348.3120	31.6647	
Total	18	938.9610		

SE(d) for comparing means of treatments not involving missing values

$$= \sqrt{\frac{2(31.6647)}{4}}$$

$$= 3.9790$$

SE(d) for comparing means of the treatment involving missing value with any other treatment

$$= \sqrt{\frac{31.6647}{4}\left[2+\frac{5}{3 \times 4}\right]}$$

$$= \sqrt{19.1310} = 4.3739$$

Example

The results in Tables are from the experiment conducted in RBD for comparing fodder yield of seven sorghum varieties.

Fodder yield t / ha

Variety	*Replication*			*Total*
	I	*II*	*III*	
1	14.5	14.0	14.0	42.5
2	16.5	16.9	16.7	50.1
3	X	16.7	17.4	34.1
4	17.6	16.9	17.5	52.0
5	18.5	17.9	17.6	54.0
6	19.3	18.3	18.8	56.4
7	19.5	19.0	X	38.5
Total	105.9	119.7	102.0	327.6

In the above data the values for variety 3 in replication I and for variety 7 in replication III are missing. Let them be denoted by X_{31} and X_{73}, respectively. They are estimated as follows: First Cycle estimates:

$$X_{31} = \left[\frac{105.9}{6}+\frac{34.1}{2}\right]/2$$

$$= \frac{17.65+17.05}{2} = 17.4$$

$$G' = 327.6+17.4 = 345.0$$

$$X_{73} = \frac{rR'+tT'-G'}{(r-1)(t-1)} = \left[\frac{rR'+tT'}{(r-1)(t-1)} - \frac{G'}{(r-1)(t-1)}\right]$$

$$= \left[\frac{3(102.0)+7(38.5)}{2 \times 6} - \frac{345.0}{2 \times 6}\right]$$

$$= \left[\frac{575.5}{12} - \frac{345.0}{12}\right]$$

$$= 47.96 - 28.75 = 19.2$$

Second cycle estimates: Substituting the value of X_{73}, we get

$$G' = 327.6 + 19.2 = 346.8$$

$$X_{31} = \frac{3(105.9)+7(34.1)}{12} - \frac{346.8}{12}$$

$$= 46.37 - 28.90 = 17.5$$

$$G' = 327.6 + 17.5 = 345.1$$

$$X_{73} = 47.96 - \frac{345.1}{12}$$

$$= 19.2$$

The values in the two cycles are nearly same. Hence, we take the estimates X_{31} = 17.5, and x73= 19.2 for further analysis. The replication, treatment and grand totals are then corrected with these values.

T_3 = 34.1 + 17.5 = 51.6, T_7 = 38.5 +19.2 = 57.7

Replication I = 105.9 + 17.5 = 123.4

Replication III = 102.0 + 19.2 = 121.2

Grand Total = 327.6 + 17.5 + 19.2 = 364.3

With these values we get,

CF = 6319.7376

Total *SS* = 6371.65- *CF* = 51.9124

Replication *SS* = 6320.7271 - *CF* = 0.9895

Treatment *SS* = 6369.69 - *CF* = 49.9524

Error *SS* = 0.9705

Ignoring the treatment classification and analysing the results in Tableas CRD, we get

CF = 5648.5137

Total *SS* = 48.2463

Replication *SS* = 5650.005 - *CF*

= 1.4913

Error *SS* = 46.755

The corrected treatment sum of squares is given as

TSS = *ESS* for CRD - ESS for RBD

= 46.7550 - 0.9705 = 45.7845

Corrected error *df.* = 12-2=10

These results are presented in the following table.

Analysis of variance for the data in Table

Sources of variation	*df*	*SS*	*MS*	*F*
Replication	2	0.9895	0.4948	5.101
Treatment	6	45.7845	7.6308	78.668**
Error	10	0.9705	0.0970	
Total	18	47.7445		

SE(d) for comparing the means of treatments not involving missing values:

$$\sqrt{\frac{2EMS}{r}} = \sqrt{\frac{2(0.0970)}{3}} = 0.2543$$

$$CD = (2.228)(0.2543) = 0.5666$$

SE(d) for comparing the mean of the treatment with missing value and that of any other treatment:

$$\sqrt{\frac{EMS}{r}\left[2+\frac{t}{(r-1)(t-1)}\right]}$$

$$= \sqrt{\frac{0.0970}{3}\left[2+\frac{7}{12}\right]}$$

$$= 0.2890$$

$$CD = (2.228)(0.2890) = 0.6439$$

For comparing the means of treatments 3 and 7,

$$SE(d) = \sqrt{EMS\left(\frac{1}{r_3}+\frac{1}{r_7}\right)}$$

where, r_3 = effective number of replications for treatment 3.

= 0+1 + 5/6=11/6

r_7 = effective number of replications for treatment 7.

=5/6+1+0=11/6

Hence,

$$SE(d) = \sqrt{(0.0970(6/11+6/11)}$$

$$= 0.3253$$

$$CD = (2.228)(0.3253) = 0.7248$$

The mean values are

T_1= 14.2, T_2 = 16.7, T_3= 17.2,T_4= 17.3

T_5= 18.0, T_6 = 18.8 and T_7= 19.2

The bar chart is given as,

T_7———T_6 T_5 T_4———T_3———T_2———T_1

The conclusions can be written based on the above bar chart.

RBD Convariance Analysis

The linear model for covariance analysis for RBD is

$$Y_{ij} = \mu + t_i + r_j + b\left(X_{ij} - X\right) + e_{ij}$$

where, the symbols have the usual meaning.

The procedure of covariance analysis for RBD is similar to that for CRD. The analysis of covariance for RBD is given in Table

Analysis of covariance for RBD.

Sources of variation	*df*	*Sum of products* *YY*	*XX*	*YX*	*Adjusted values for Y* *df*	*SS*	*MS*
Total	$rt-1$	G_{yy}	G_{xx}	G_{yx}	-	-	-
Replication	r-1	R_{yy}	R_{xx}	R_{yx}	-	-	-
Treatment	r-1	T_{yy}	T_{xx}	T_{yx}	-	-	-
Error	$(r-1)(t-1)$	E_{yy}	E_{xx}	E_{yx}	$(r-1)(t-1)-1$	E'_{yy}	*EMS*
Treatment + Error	$r(t-1)$	$T_{yy}+E_{yy}$	$T_{xx}+E_{xx}$	$T_{yx}+E_{yx}$	$r(t-1)$	$(T_{yy}+E_{yy})^r$	-
Treatment adjusted for the average regression within treatments					$t-1$	T_{yy}'	*TMS*

$$E_{yy}' = E_{yy} - \frac{\left(E_{yx}\right)^2}{E_{xx}}$$

$$\left(T_{yy}+E_{yy}\right)' = \left(T_{yy}+E_{yy}\right) - \frac{\left(T_{yx}+E_{yx}\right)^2}{\left(T_{xx}+E_{xx}\right)}$$

$$T_{yy}' = \left(T_{yy}+E_{yy}\right)' - E_{yy}'$$

The adjusted treatment means and *SE(d)* are found by using the formulae given for analysis of covariance for CRD.

Example

A fertilizer trial on IR - 20 paddy was conducted in RBD. The grain yield was the primary variable, *Y*. The number of productive tillers per hill was observed as mean of ten hills and it was the covariate, *X*. The results are presented in Table

Grain yield, kg per plot and number of productive tillers per hill.

Treatment	*Replication* I		II		III		IV		*Total*	
	Y	X	Y	X	Y	X	Y	X	Y	X
Control	7.0	5.1	6.4	5.5	8.0	5.0	6.9	5.5	28.3	21.1

Contd...

Treatment	*Replication*								*Total*	
	I		II		III		IV			
AN_1	10.8	6.5	9.0	6.3	10.5	6.7	9.6	6.5	39.9	26.0
AN_2	13.0	7.6	12.6	7.6	12.0	7.3	13.0	8.6	50.6	31.1
AN_3	15.0	8.5	14.8	8.9	14.0	9.5	14.0	9.5	57.8	36.4
AN_4	14.8	10.4	15.0	9.5	13.0	9.7	14.1	10.1	56.9	39.7
UN_1	9.9	6.3	10.5	6.4	9.0	6.3	9.6	6.2	39.0	25.2
UN_2	13.1	7.5	11.9	7.1	12.9	7.8	12.5	7.9	50.4	30.3
UN_3	14.4	8.1	14.2	9.5	13.5	9.5	14.1	8.8	56.2	35.9
UN_4	15.0	9.2	14.8	10.1	13.8	10.4	12.8	9.9	56.4	39.6
Total	113.0	69.2	109.2	70.9	106.7	72.2	106.6	73.0	435.5	285.3

Analysis for *Y*:

$$CF = \frac{(435.5)^2}{36} = 5268.3403$$

$$\text{Total } SS = G_{yy} = (7.0)^2 + (10.8)^2 + \ldots\ldots + (12.8)^2 - CF$$

$$= 227.6097$$

$$\text{Replication } SS = R_{yy} = \frac{1}{9}\left[(113.0)^2 + (109.2)^2 + (106.7)^2 + (106.6)^2\right] - CF$$

$$= 3.0030$$

$$\text{Treatment } SS = T_{yy} = \frac{1}{4}\left[(28.3)^2 + (39.9)^2 + \ldots\ldots + (56.4)^2\right] - CF$$

$$= 214.7272$$

$$\text{Error } SS = E_{yy} = 9.8795$$

Analysis for *Y and X*:

$$CF = \frac{(435.5)(285.3)}{36} = 3451.3375$$

$$\text{Total sum of products} = total\,SP = G_{yx}$$

$$= \left[(7.0)(5.1) + (10.8)(6.5) + \ldots\ldots + (12.8)(9.9)\right] - CF$$

$$= 3582.1000 - 3451.3375 = 130.7625$$

Replication $SP = R_{yx}$

$$= \frac{1}{9}\left[(113.0)(69.2) + (109.2)(70.9) + ... + (106.6)(73.0)\right] - CF$$

$$3449.7133 - CF = 1.6242$$

Treatment $SP \quad = T_{yx}$

$$= \frac{1}{4}\left[(28.3)(21.1) + (39.9)(26.0) + ... + (56.4)(39.6)\right] - CF$$

$$= 3582.9950 - CF = 131.6575$$

Error $SP = E_{yx} = 0.7292$

For the covariate X,

$TMS = 88.8900/8 = 11.1112$

$EMS = 4.0789/24 = 0.1700$

$$F = \frac{11.1112}{0.1700} = 65.36$$

Since F is significant at 1 % level of significance, we conclude that the covariate is also affected by the treatments. The final conclusion has to be drawn keeping this in view.

$$b \quad = \frac{E_{yx}}{E_{xx}} = \frac{0.7292}{40789} = 0.1788$$

$$F(\text{for } b) \quad = \left[\frac{E_{yx}^2}{E_{xx}}/1\right] \div \left[\left\{E_{yy} - \frac{E_{yx}^2}{E_{xx}}\right\}/(r-1)(t-1)-1\right]$$

$$\frac{E_{yx}^2}{E_{xx}} = \frac{(0.7292)^2}{4.0789} = 1.3036$$

$$E_{yy} - \frac{E^2{}_{yx}}{E_{xx}} = 9.8795 - 1.3036 = 8.5859$$

$$(r-1)(t-1) - 1 = 24 - 1 = 23$$

For 1 and 23 *df and* $\alpha = 0.05$, the table value of F is 4.28. In this case F is not significant and hence b is not significant. Since b is not significant, the effect covariate in reducing the error will not be significant. However, for illustration purpose complete analysis is shown in Table.

Analysis of covariance for the data in Table

Sources of variation	df	Sum of products YY	XX	YX	Adjusted values for Y df	SS	MS	F
Total	35	227.6097	93.8875	130.7625	—			
Replication	3	3.0030	0.9186	-1.6242	—			
Treatment	8	214.7272	88.8900	131.6575	—			
Error	24	9.8795	4.0789	0.7292	23	8.5759	0.3729	
Treatment + Error	32	224.6067	92.9689	132.3867	—	—	—	
Treatment adjusted treatments for the average regression within					8	27.5136	3.4392	9.223**

$$E'_{yy} = E_{yy} - \frac{E_{yx}^2}{E_{xx}} = 9.8795 - \frac{(0.7292)^2}{4.0789} = 8.5759$$

$$\left(T_{yy} + E_{yy}\right)' = \left(T_{yy} + E_{yy}\right) - \left[\left(T_{yx} + E_{yx}^2\right) / \left(T_{xx} + E_{xx}\right)\right]$$

$$= 224.6067 - \frac{(132.3867)^2}{92.9689}$$

$$= 36.0895$$

$$T_{yy}' = \left(T_{yy} + E_{yy}\right)' - E_{yy}'$$

$$= 36.0895 - 8.5759 = 27.5136$$

Computation of adjusted treatment means:

Treatment	Y_i	X_i	$b(\bar{x}_i - \bar{x}) = 0.1788(\bar{x}_i - 7.93)$	$\bar{y}_i' = [y_i - b - b(\bar{x} - \bar{x})]$
1	7.08	5.28	- 0.4738	7.55
2	9.98	6.50	-0.2557	10.24

Contd:..

Treatment	Y_i	X_i	$b(\bar{x}_i - \bar{x}) = 0.1788(\bar{x}_i - 7.93)$	$\bar{y}'_i = [y_i - b - b(\bar{x} - \bar{x})]$
3	12.65	7.78	-0.0268	12.68
4	14.45	9.10	0.2092	14.24
5	14.42	9.92	0.3558	14.06
6	9.75	6.30	-0.2914	10.04
7	12.60	7.58	- 0.0626	12.66
8	14.05	8.98	0.1877	13.86
9	14.10	9.90	0.3522	13.75

$$SE(d) = \sqrt{\frac{2EMS}{r}\left[1+\frac{T_{xx}}{(-1)E_{xx}}\right]}$$

$$= \sqrt{\frac{2(0.3729)}{4}\left[1+\frac{88.8900}{8(4.0789)}\right]}$$

$$= \sqrt{(0.18645)(3.7241)}$$

$$= 0.8333$$

$$CD = (2.069)(0.8333) = 1.724$$

Covariance Analysis with two or more Covariates

The analysis of covariance technique may be extended to two or more covariates.

In this section the analysis of covariance with two covariates used with RBD is described. Suppose that there are two covariates, X and Z.

The variate values have to be adjusted for the linear effects of both X and Z through the use of multiple regression equation. The structure of analysis will be as in Table.

The adjusted error sum of squares for Y is given by,

$$E_{yy}' = E_{yy} - b_x E_{yx} - b_z E_{yz}$$

where, $$b_x = \frac{\left(E_{zz}E_{yx} - E_{yz}E_{xz}\right)}{\left(E_{xx}E_{zz} - E_{xz}^2\right)}$$

$$b_z = \frac{\left(E_{xx}E_{yz} - E_{yx}E_{xz}\right)}{\left(E_{xx}E_{zz} - E_{xz}^2\right)}$$

The adjusted treatment sum of squares is given by

$$T_{yy}' = (T+E)'_{yy} - E_{yy}'$$

where, $$(T+E)_{yy}' = \left(T_{yy} + E_{yy}\right) - \frac{\left(T_{yx} + E_{yx}\right)^2}{T_{xx} + E_{xx}} - \frac{\left(T_{yz} + E_{yz}\right)^2}{T_{zz} + E_{zz}}$$

The adjusted mean for i[th] treatment is given by,

$$Y_i' = Y_i - b_x(X_i - X) - b_z(Z_i - Z)$$

The standard error for the difference between two adjusted treatment means is given by

$$SE(d) = \sqrt{\frac{2EMS}{r}\left[1 + \frac{T_{xx}E_{zz} - 2T_{xz}E_{xz} + T_{zz}E_{xx}}{(t-1)\left(E_{xx}E_{zz} - E_{xz}^2\right)}\right]}$$

In case of missing values, the standard error for the difference between the mean of a treatment with no missing value and that of a treatment with missing value is given by

$$SE(d) = \sqrt{\frac{EMS}{r}\left[2 + \frac{E_{zz}}{r\left(E_{xx}E_{zz} - E_{xz}^2\right)}\right]}$$

For comparing the means of the treatments with missing values, we have

Analysis of Covariance for RBD with two Covariates.

Sources of Variation		*Sum of Products*						*Adjusted Values for Y*			
	df	*YY*	*XX*	*ZZ*	*YX*	*YZ*	*XZ*	*df*	*SS*	*MS*	*F*
Total	*rt*-1	G_{yy}	G_{xx}	G_{zz}	G_{yx}	G_{yz}	G_{xz}	—	—	—	—
Replication	*r*-1	R_{yy}	R_{xx}	R_{zz}	R_{yx}	R_{yz}	R_{xz}	—	—	—	—
Treatment	*t*-1	T_{yy}	T_{xx}	T_{zz}	T_{yx}	T_{yz}	T_{xz}	–	–	—	—
Error	(r-1)(t-1)	E_{yy}	E_{xx}	E_{zz}	E_{yx}	E_{yz}	E_{xz}	(*r*-1)(*t*-1)-2	E_{yy}'	*EMS*	—
Treatment + error	—	$T_{yy} + E_{yy}$	$T_{xx} + E_{xx}$	$T_{zz} + E_{zz}$	$T_{yx} + E_{yx}$	$T_{yz} + E_{yz}$	$T_{xz} + E_{xz}$	—	—	—	—
Treatment adjusted for the average regression with treatments								*t* - 1	T_{yy}'	*TMS*	$\frac{TMS}{EMS}$

$$SE(d) = \sqrt{\frac{2EMS}{r}\left[1+\frac{\left(E_{xx}E_{zz}+2E_{xz}\right)}{r\left(E_{xx}E_{zz}-E_{xz}^{2}\right)}\right]}$$

Example: Consider the data in Example in which two values are missing. The data arrangement with two covariates will be as in Table

Arrangement of RBD data in Table, incorporating covariates

Treatment	*Replication*									*Total*		
	I			II			III					
	Y	*X*	*Z*	*Y*	*X*	*Z*	*Y*	*X*	*Z*	*Y*	*X*	*Z*
1	14.5	0	0	14.0	0	0	14.0	0	0	42.5	0	0
2	16.5	0	0	16.9	0	0	16.7	0	0	50.1	0	0
3	0	1	0	16.7	0	0	17.4	0	0	34.1	1	0
4	17.6	0	0	16.9	0	0	17.5	0	0	52.0	0	0
5	18.5	0	0	17.9	0	0	17.6	0	0	54.0	0	0
6	19.3	0	0	18.3	0	0	18.8	0	0	56.4	0	0
7	19.5	0	0	19.0	0	0	0	0	1	38.5	0	1
Total	105.9	1	0	119.7	0	0	102.0	0	1	327.6	1	1

The computation of sums of squares and sums of products are carried out as in analysis of covariance with one covariate. The results are presented in Table.

Computation of Sums of Squares and Sums of Products.

	YY	XX	ZZ	YX	YZ	XZ
CF	5110.5600	0.0476	0.0476	15.6000	15.6000	0.0476
Total	586.2000	0.9524	0.9524	- 15.6000	- 15.6000	- 0.0476
Rep.	24.7114	0.0953	0.0953	- 0.4714	- 1.0286	- 0.0476
Treat.	143.5333	0.2857	0.2857	- 4.2333	- 2.7667	- 0.0476
Error	417.9553	0.5714	0.5714	- 10.8953	- 11.8047	0.0476
Treat + Error	561.4886	0.8571	0.8571	-15.1286	-14.5714	0

$$b_x = \frac{E_{zz}E_{yx} - E_{yz}E_{xz}}{E_{xx}E_{zz} - E_{xz}^2}$$

$$= \frac{(0.5714)(-10.8953)-(-11.8047)(0.0476)}{(0.5714)(0.5714)-(0.0476)^2}$$

$$= \frac{-5.663671}{0.324232} = -17.4680$$

$$b_z = \frac{E_{xx}E_{yz} - E_{yx}E_{xz}}{E_{xx}E_{zz} - E_{xz}^2}$$

$$= \frac{(0.5714)(-11.8047)-(-10.8953)(0.0476)}{0.324232}$$

$$= \frac{-6.226589}{0.324232} = -19.2041$$

$$E_{yy}' = E_{yy} - b_x E_{yx} - b_z E_{yz}$$

$$= 417.9553-(-17.4680)(-10.8953)-(-19.2041)(-11.8047)$$

$$= 417.9553-190.1391-226.6986$$

$$= 0.9376$$

$$(T+E)_{yy}{}' = \left(T_{yy}+E_{yy}\right) - \left[\frac{\left(T_{yx}+E_{yx}\right)^2}{T_{xx}+E_{xx}}\right] - \left[\frac{\left(T_{yz}+E_{yz}\right)^2}{T_{zz}+E_{zz}}\right]$$

$$= 561.4886 - \frac{(-15.1286)^2}{0.8571} - \frac{(-14.5714)^2}{0.8571}$$

$$= 561.4886 - 267.0336 - 247.7257$$

$$= 46.7293$$

$$T_{yy}{}' = (T+E)_{yy}{}' - E_{yy}$$

$$= 46.7293 - 0.9376 = 45.7917$$

The above results are presented in Table.

Analysis of variance for the data in Table.

Sources of variation	*d.f.*	*SS*	*MS*	*F*
Treatment	6	45.7917	7.6320	81.365**
Error	10	0.9376	0.0938	

The standard error for the difference between the means of treatments not involving missing values is

$$SE(d) = \sqrt{\frac{2EMS}{r}\left[1+\frac{T_{xx}E_{zz}-2T_{xz}E_{xz}+T_{zz}E_{xx}}{(t-1)\left(E_{xx}E_{zz}-E_{xz}^2\right)}\right]}$$

$$T_{xx}E_{zz} - 2T_{xz}E_{xz} + T_{zz}E_{xx} = (0.2857)(0.5714) - 2(-0.0476)(0.0476) + (0.2857)(0.5714)$$

$$= 0.331029$$

$$(t-1)\left(E_{xx}E_{zz}-E_{xz}^2\right) = 6(0.5714)(0.5714) - (0.0476)^2$$

$$= 6(0.324232) = 1.945392$$

$$SE(d) = \sqrt{\frac{2(0.0938)}{3}\left[1+\frac{0.331029}{1.945392}\right]}$$

$$= \sqrt{(0.062533)(1.170160)}$$

$$= 0.2705$$

$$CD = (2.228)(0.2705) = 0.6027$$

For comparing the mean of a treatment not involving missing value with that of a treatment involving missing value,

$$SE(d) = \sqrt{\frac{EMS}{r}\left[2+\frac{E_{zz}}{r\left(E_{xx}E_{zz}-E_{xz}^2\right)}\right]}$$

$$= \sqrt{\frac{0.0938}{3}\left[2+\frac{0.5714}{3(0.324232)}\right]}$$

$$= 0.2844$$

$$CD = (2.228)(0.2844) = 0.6336$$

For comparing the means of treatments 3 and 7,

$$SE(d) = \sqrt{\frac{2EMS}{r}\left[1+\frac{E_{xx}+E_{zz}+2E_{xz}}{r\left(E_{xx}E_{zz}-E_{xz}^2\right)}\right]}$$

$$= \sqrt{\frac{2(0.0938)}{3}\left[1+\left\{\frac{0.05714)+0.5714+2(0.0476)}{3(0.324232)}\right\}\right]}$$

$$= 0.3770$$

$$CD = (2.228)(0.3770) = 0.8400$$

The adjusted mean values are calculated as follows:

Treatment	$\bar{y}$	$b_x(\bar{x}_i-\bar{x})$	$b_z(\bar{z}_i-\bar{z})$	*Adjusted mean* $\bar{y}$ = (2)- (3)- (4)
(1)	(2)	(3)	(4)	
1	14.2	0.8315	0.9141	12.45
2	16.7	0.8315	0.9141	14.95
3	11.4	- 4.9906	0.9141	15.48
4	17.3	0.8315	0.9141	15.55
5	18.0	0.8315	0.9141	16.25
6	18.8	0.8315	0.9141	17.05
7	12.8	0.8315	- 5.4674	17.44

BIBLIOGRAPHY

Adam, N. : *Encyclopaedia of Plant Physiology*, Springer Verlag, Berlin, 1975.

Alam, B. : *Advanced Plant Physiology*, Longman, London, 1984.

Anderson, Edgar : *Plants, Man and Life*, University of California Press, California, 1967.

Arvill, R. : *Man and Environment : Crisis and the Strategy of Choice*, Penguin, New York, 1967.

Bailey, L. H. : *Hand Book of the Multiplication of Plants*, Discovery Publishing House, New Delhi, 1999.

Beevers, L. : *Nitrogen Metabolism in Plants*, Edward Arnold, London, 1976.

Bhatia, B. M. : *Poverty, Agriculture and Economic Growth*, Vikas Publishing House, New Delhi, 1977.

Billihgs, W. D. : *Plant, Man and the Ecosystem*, MacMillan, London, 1971.

Bowling, D. J. F. : *Uptake of Ions by Plant Roots*, Chapmann and Hall, London, 1976.

Bray, C. : *Nitrogen Metabolism in Plants*, Longman, London, 1984.

Candolle, Alphonse De : *Origin of Cultivated Plants*, Hafner, New York, 1886.

Crafts, A. S. and C. E. Crisp, : *Phloem Transport in Plants*, W.H. Freeman, San Francisco, 1973.

Dadiani, N. K. : *The Survey of Indian Agriculture,* Kasturi & Sons Ltd., Chennai, 1999

Deckock, P. C. : *Mineral Nutrition of Plants : a Symposium,* University of Oxford, New York, 2002.

Dennis, D. T. and Turpin, D. H. : *Plant Physiology, Biochemistry and Molecular Biology,* Longman, London, 1990.

Douglas, J. S. : *Advanced Guide to Hydroponics,* Pelham Books, London, 1976.

Dubey, R. S. : *Agricultural : Issues and Application,* Gyan Books, New Delhi, 1987.

Dunn, Edgar S. : *Location of Agricultural Production,* University of Florida Press, New York, 1954.

Evans, L. T. : *Day Length and the Flowering of Plants,* W.A. Benjamin, California, 1975.

Farmer, J. B. : *Study of Plant Life,* Discovery Publishing House, New Delhi, 1993.

Frankel, F. R. : *India's Green Revolution : Agriculture Costs of Land Growth,* Princeton University Press, Princeton, 1971.

Ftzler M. E. : *Plant Lectins : Molecular and Biological Aspects,* Academic Press, New York, 1985.

Gangawane, L. V. : *Vistas in Mycology and Plant Pathology,* Commonwealth Publishers, New Delhi, 2000.

Gauch, H. G. : *Inorganic Plant Nutrition,* Hutchinson and Ross, London, 1972.

Green, J. R. : *An Introduction to Vegetable Physiology,* MacMillan, London, 1911.

Gregor, Howard F. : *Geography of Agriculture : Themes in Research,* Prentice Hall, New Jersey, 1970.

Gupta, G.P. : *Test Book of Plant Diseases,* Discovery Publishing House, New Delhi, 2001.

Hansen, E. : *Post Harvest Physiology of Fruits,* Praeger Publishers, New York, 1966.

Hewitt, E. J. and Cutting, C. V. : *Nitrogen Assimilation of Plants,* Academic Press, London, 1982.

Hussain, M. : *Systematic Agricultural Geography,* Rawat Publications, Jaipur, 1999.

Isser, G. P. : *Changing Face of Poverty and Globalisation,* Gyan Books, New Delhi, 2002.

James, A.: *Weather and Agriculture,* Pergamon, Oxford, 1967.

Joshi, H. L. : *Agricultural of India,* Rawat Publications, Jaipur, 1990.

Karanth, Apoorva : *Plant Genetics and Food Crop Science,* Dominant Books, New Delhi, 1999.

Kirkby, E. A. : *Nitrogen Nutrition of Plants,* University of Leeds, Leeds, 1970.

Kramer, A. L. : *Principles of Biochemistry,* Worth Publishers, New York, 1982

Kramer, P. J. : *Plant and Soil Water Relationship : A Modern Synthesis,* McGraw Hill Book Company, New York, 1969.

Kumar, Arun : *Agricultural Development : Today and Tomorrow,* Anmol Publications, New Delhi, 1995.

Kumar, T. : *History of Rice in India : Mythology, Culture and Agricultural,* Gyan Books, New Delhi, 1988.

Laxmi Narasaiah, M. : *Agricultural Production,* Discovery Publishing House, New Delhi, 1999.

Mazliak, P. : *Lipid Metabolism in Plants,* Chapmann and Hall, London, 1973.

McKee, H. S. : *Nitrogen Metabolism in Plants,* Oxford University Press, London, 1962.

McNeil, M. Darvill : *Structure and Function of the Primary Cell Walls of Plants,* McGraw, London, 1984.

McRae, S. C. and Burnham, C. P. : *Land Evaluation,* Oxford University Press, New York, 1981.

Meidner, H. and Sheriff, D. W. : *Water and Plants,* Blackie and Sons, Glasgow, 1976.

Mengel, K. and Kirkby, E.A. : *Principles of Plant Nutrition*, Praeger Publishers, New York, 1974.

Mishra, S. R. : *Bacterial Plant Diseases*, Discovery Publishing House, New Delhi, 2003.

Misra, B., Kar, G. C. and Misra, S. N. : *Agro-Industries and Economic Development*, Deep & Deep Publications, New Delhi, 2002.

Mitter, V. : *Growth of Urban Informal Sector*, Deep & Deep Publications, New Delhi, 1988.

Mukherji, Shekhar : *Land-Use Change, Environment and Sustainable Development*, Banaras Hindu University, Varanasi, 2000.

Nomura, M. : *The Control of Ribosome Synthesis*, Edward Arnold, London, 1986.

Novikoff, A. B. : *The Endoplasmic Reticulum : a Cytochemist's View*, Oxford, New York, 1976.

Numa, S. : *Fatty Acid Metabolism and its Regulation*, Janis, London, 1984.

Nutman, P. S. : *Symbiotic Nitrogen Fixation in Plants*, Cambridge University Press, Cambridge, 1976.

Ogawa, Y. and R. W. King : *Plant and Cell Physiology*, Springer-Verlag, Berlin, 1990.

Opik, H. : *The Respiration of Higher Plants*, Edward Arnold Publishers, London, 1980.

Palladin, V. I. : *Plant Physiology*, Arihant Publishers, Jaipur, 1988.

Pannikkar, K. M. : *India and the Indian Ocean : An Essay on the Influence of Sea Power on Indian History*, George Allen and Unwin, London, 1951.

Ramshaw, J. A. M. : *Structure of Plant Proteins*, Oxford University Press, New York, 1982.

Robert, W. : *Middle America : Its Land and Peoples*, Prentice Hall, Englewood Cliff, 1989.

Roberts, J. A. and Tucker, G. A. : *Ethylene and Plant Development*, Butterworth, London, 2001.

Rubenstein, J. M., and Bacon, R. S. : *The Cultural Landscape : An Introduction to Human Geography*, Prentice Hall, New Delhi, 1990.

Salisbury, F. B. and C. W. Ross, : *Plant Physiology*, Woodworth Publishing Co., California, 1967.

Samanta, R. K. : *Agricultural Estension in Changing World Perspective*, Uppal Publishing House, New Delhi, 1995.

Shandhya, T. K. : *Agricultural Credit and Nabard*, Deep & Deep Publications, New Delhi, 2003.

Sharma, D. N. : *Minerals of India*, National Book Trust, New Delhi, 1994.

Sharma, S. S. : *Sustainable Agriculture : Poverty and Food Security*, Rawat Publications, Jaipur, 2002.

Sharma, V. K. : *Trees and Protection of Environment*, Deep & Deep Publications, New Delhi, 2004.

Siegelman, H. W. and G. Hind : *Photosynthetic Carbon Assimilation*, Plenum, New York, 1978.

Singh, D. K. : *Technological Change and Agricultural Development*, Deep & Deep Publications, New Delhi, 1993.

Singh, Harjinder : *Agricultural Problems in India*, Gyan Books, New Delhi, 1988.

Singh, L. : *Green Revolution and Cropping Pattern*, Deep & Deep Publications, New Delhi, 1993.

Singh, Surendra : *Agricultural Development in India : A Regional Analysis*, Kaushal Publications, Shillong, 1994.

Slatyer, R. O. : *Plant Water Relationship*, Academic Press, New York, 1967.

Somani, K. K. : *Indian Economics and Social Traditions*, Gyan Books, New Delhi, 2002.

Srivastava O. S. : *Agricultural Economics*, Rawat Publications, Jaipur, New Delhi, 1996.

Steward, F. C. : *Plant Physiology, A Treatise*, Academic Press, New York, 1963.

Street, H. E. and H. Opik : *The Physiology of Flowering Plants*, Edward Arnold, London, 1970.

Subrahmanya, S. : *Studies in Agricultural Development*, Deep & Deep Publications, New Delhi, 1996.

Sudhir, M. : *Applied Biotechnology and Plant Genetics*, Dominant Books, New Delhi, 2000.

Thomas, C. : *A History of the Land and Agriculture*, Prentice Hall, Englewood Cliff, 1978.

Valcarcel, M. : *Weather and Agriculture*, Oxford University Press, London, 1997.

Varner, J. E., Flint D. and R. Mitra, : *Characterization of Protein Metabolism in Cereal Grains*, Academic Press, New York, 1976.

Venkateshwarlu : *Developing Agricultural Technology*, Rawat Publications, Jaipur, 1998.

Verma, M. : *The Survey of Indian Agriculture*, Kasturi & Sons, New Delhi, 1999.

Verma, P. C. : *Surplus Manpower in Agriculture and Employment Policy*, Deep & Deep Publications, New Delhi, 1991.

Vyas, V. S. : *Policies for Agricultural Development*, Rawat Publications, Jaipur, 1997.

Wilkins, M. B. : *The Physiology of Plant Growth and Development*, McGraw Hill book, London, 1969.

Zeevart, J. A. D. : *Physiology of Flower Formation*, McGraw Hill, London, 1976.

Zimmermann, M. H. : *Encyclopeadia of Plant Physiology*, Academic Press, New Work, 1982.

INDEX

G

H

I

J

L

M

N

O

P

Q

R

S

T

U

V

W

Y

Z

❑❑❑